OBSERVING FOR THE FUN OF IT

MELANIE MELTON

FROM THE PUBLISHERS OF ASTRONOMY MAGAZINE

 Published by Kalmbach Publishing Co., 21027 Crossroads Circle, P. O. Box 1612, Waukesha, WI 53187.

Printed in the United States of America

Publisher's Cataloging-in-Publication Data
(Prepared by Quality Books, Inc.)

Melton, Melanie.
Observing for the fun of it / Melanie Melton.
— Waukesha, WI : Kalmbach Pub., 1996.
p. cm.

ISBN 0-913135-26-7

1. Astronomy—Popular works. I. Title.

QB44.2.M45 1996 520
QBI95-20339

Star Maps: Software Bisque
Art Director: Kristi Ludwig
Illustrations: Sallie Spencer
Book Design: Mark Watson

ACKNOWLEDGMENTS

The proofreaders: Becky Cooper, Nancy Luis, and Greg Novacek (my friends at LAPO), Leslie Willett, Corkie Martin, and Sue Peterson. Thanks, guys! Steve Golden—for all the moral support. Ralph Roncoli—for introducing me to Sallie. Tom and Steve from Software Bisque—for the last-minute major help with the star maps. Bob and Helen Melton—my parents.

Dedicated to my friends at Lake Afton Public Observatory, Wichita, Kansas.

Contents

Introduction

Have you ever wondered if you could see the total lunar eclipse you heard about on television? What about that meteor shower you read about in the paper? Have you ever tried to find a constellation using a star chart, or wanted to buy a telescope for yourself or your grand-kids and walked out of a store more confused and perplexed than when you went in?

If any of these experiences sound familiar, you are not alone. Many people share a fascination with the stars, planets, and Moon we see in the night sky. Fortunately, you don't have to devote enormous amounts of time to enjoy the night sky. All that is really required are a few observing tricks, a couple of comfort items, and an occasional hour or two away from your television at night. In no time at all, you can enjoy watching meteors burning up as they enter Earth's atmosphere or show your friends how to find a couple of your favorite constellations. Observational astronomy can be as light-hearted or as serious as you wish. After all, the night sky is big enough for everyone.

OBSERVING FOR THE FUN OF IT

PART I

OBSERVING THE SKY WITH ONLY YOUR EYES

1

CONSTELLATIONS

The *Big Dipper, Orion,* the Hunter, *Leo,* the Lion, *Cygnus,* the Swan, *Delphinus,* the Dolphin. Do any of these sound familiar? How about *Gemini, Cancer,* or *Pegasus* and *Perseus*? With the exception of one, all of these are constellations found in our night skies. For centuries, the constellations have lit the way home for many, risen as signals to start the planting or to bring in the harvest for others. Their stories have entertained and terrified listeners around many campfires. Today, with the advent of calendars, compasses, and coordinate systems, constellations are not used as much as they used to be. However, they are still great for telling stories around a campfire or to give you a sense of direction in the dark.

There are many other constellations besides the ones listed above. Eighty-eight, to be exact. You may be familiar with some and don't even know it. What about *Taurus, Aries, Virgo,* and *Sagittarius*? These and other signs of the zodiac got their start in the sky. This chapter will explain not only how the zodiac signs found their way into the stars, but also will give you a better understanding of what constellations are and some tricks to help you find them.

Before we get started, though, let's go back to the first paragraph. Which one of the names listed is not a constellation? I hate to disappoint you, but it's the Big Dipper. Although one of the most familiar and easily identified group of stars in the sky, the Big Dipper is just a nickname given to seven stars in the constellation *Ursa Major,* the big bear.

Don't get upset, though. Even if the Big Dipper isn't an official constellation, you can still use it to help find your way to other official constellations. Before offering some tricks to identify those other constellations, let's talk about what an official constellation is.

INTERSTELLAR CONNECT-THE-DOTS

Have you ever gone outside on a summer day and seen shapes in the clouds? A dragon, perhaps? Or a sailboat? With a little imagination, it's easy to see all kinds of shapes within the clouds. But those shapes can change rapidly. In one ten-minute setting, a dinosaur can become a tree and then a fish, depending on how the winds blow and how active your imagination is.

The constellations we know today are the result of a nighttime version of the cloud game. Long ago, people imagined they saw characters from their myths and legends among the stars. Since the stars don't move like clouds, these shapes have remained constant through the years and have been passed down from generation to generation.

Not every country had the same constellations. That makes sense if you think about it. After all, different cultures have different stories and heroes. Most of the 88 constellations we know today came from the ancient Greeks and Romans. If we were going to name them today, we would probably use our own modern-day heroes such as Superman, Indiana Jones, or maybe even Tyrannosaurus Rex.

Imagination plays an important role in trying to find a constellation. You may have noticed this already if you have tried to find any on your own. In fact, imagination is the only thing holding a constellation together. The stars in a constellation only appear to be held together as a physical group. In reality, each star has very little to do with its celestial neighbor.

Constellation Facts

There are 88 constellations in the sky. They can be divided into categories in different ways: the constellations of the Northern Hemisphere and the Southern Hemisphere, for instance; or spring, summer, fall, and winter constellations.

However you list them, the fact is you can't see all of them from any given location. If you happen to live in the Northern Hemisphere you can't see some of the constellations in the Southern Hemisphere because they are below your horizon. It works the same way for people who live in the Southern Hemisphere. People in Australia or South America can never see the North Star.

If you are like most people and stuck in one hemisphere, look at the bright side. There's no need to learn all 88 constellations! Instead, you can stick with the brightest constellations of each season that are seen from your own hemisphere.

Why do we have different constellations for different seasons? Imagine that you go outside and brave the cold of a long winter night. You see such grand winter constellations as *Orion,* the Hunter, and *Taurus,* the Bull. The summer constellations are nowhere to be seen.

Where are they? you may ask. Are they off warming their toes and sipping hot chocolate in some cozy, warm lodge? No, *Cygnus,* the Swan, *Lyra,* the Harp, and all the other summer constellations are shining just as they do in the summer months, just as they do 365 days of the year. It's not their fault we can't see them. It's our fault. (Actually, it's the fault of the Sun and Earth, to be precise.)

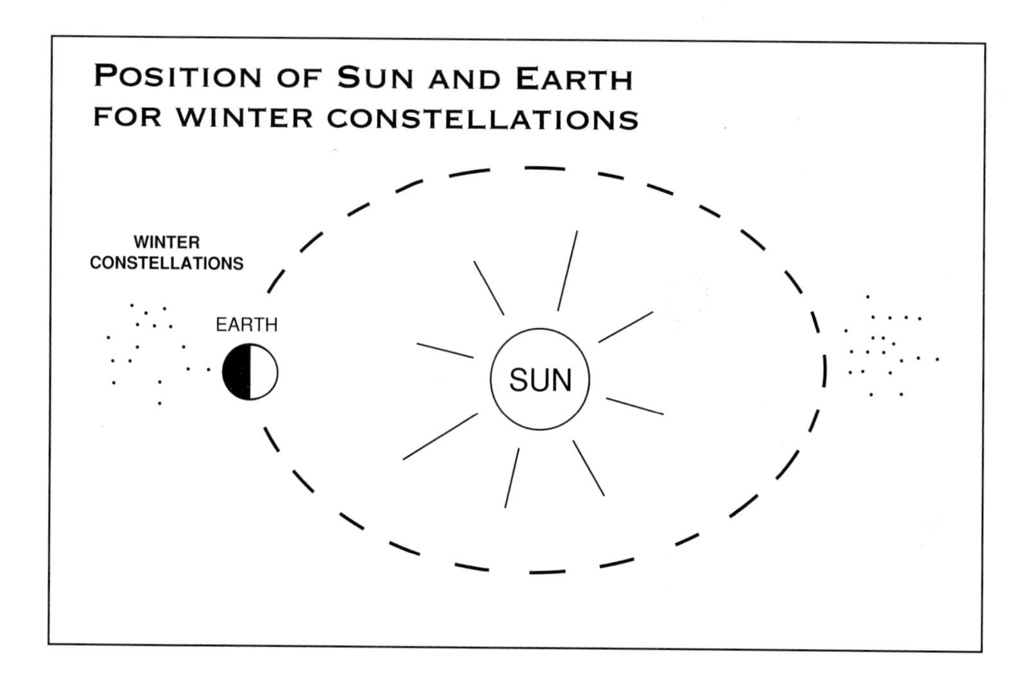

During winter in the Northern Hemisphere, as our side of Earth rotates into night, we see *Orion, Taurus,* and the other winter constellations. The summer constellations of *Cygnus* and *Lyra* are blocked by the Sun in the daytime skies, so they can't be seen. The Sun's light drowns them out.

Six months later, Earth has moved around in its orbit and the season in the Northern Hemisphere is summer. During a summer night, we can now see *Cygnus* and *Lyra,* with *Orion* and *Taurus* now hidden by the Sun's light during the day.

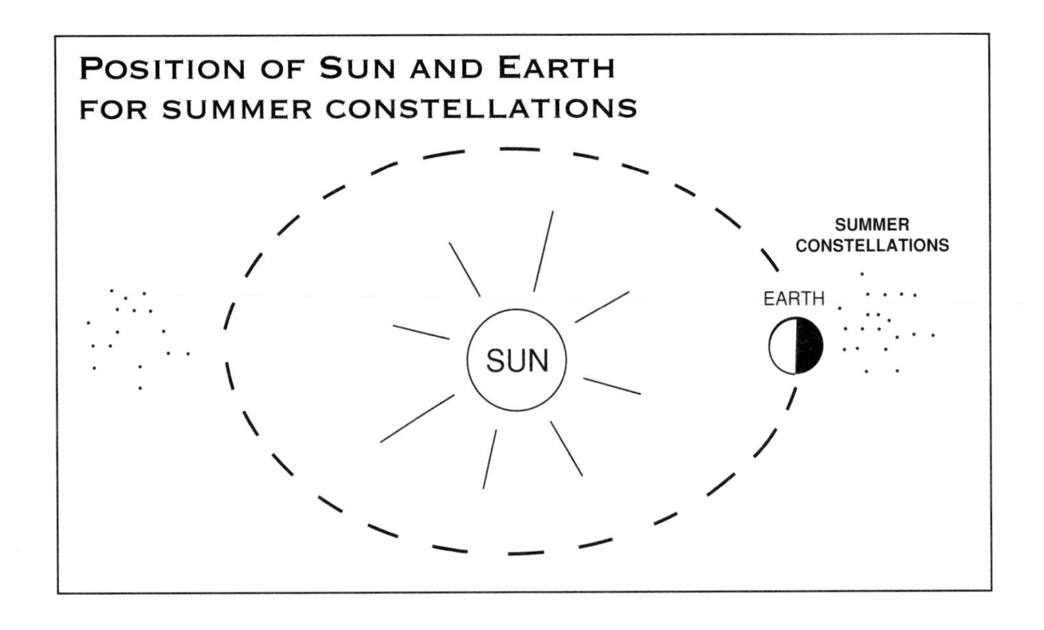

A Few Constellations by the Season

(With a Northern Hemisphere Slant)

Here's a list of some of the brighter constellations that can be seen from the Northern Hemisphere. They are listed by season, when the constellation can be seen in the sky just after sunset.

Spring	Summer	Fall	Winter
Bootes	Aquila	Andromeda	Aries
Cancer	Corona Borealis	Aquarius	Auriga
Canes Venatici	Cygnus	Camelopardalis	Canis Major
Coma Berenices	Delphinus	Capricornus	Canis Minor
Corvus	Hercules	Cassiopeia	Gemini
Crater	Libra	Cepheus	Lepus
Hydra	Lyra	Cetus	Orion
Leo	Ophiuchus	Draco	Taurus
Leo Minor	Sagitta	Pegasus	
Virgo	Sagittarius	Perseus	
	Scorpius	Pisces	
	Vulpecula	Triangulum	

CONSTELLATIONS OF THE ZODIAC

Now that you understand all about different constellations in different seasons and different hemispheres, where do Leo, Cancer, Gemini, and the other constellations of the zodiac fit in? Well, they fit in with the other constellations, but with a unique twist.

That twist was first noticed a couple of thousand years ago by the ancient Babylonians. Astronomers of that day (or night, to be exact) noticed that when the Sun set each month and the sky grew dark, a different constellation would appear where the Sun had just disappeared.

If things were constant in the universe, the same stars should always be visible just after the Sun goes down. But as the astronomers watched, they soon realized that this was not the case. Instead, it was as if the Sun were moving through the stars, each month setting in a different constellation. The movement wasn't completely random, however. Every year, the Sun revisited the same star patterns. The Babylonians divided the Sun's path into 12 constellations: Aries, Taurus, Gemini, Cancer, Leo, Virgo, Libra, Scorpius, Sagittarius, Capricornus, Aquarius, and Pisces—the constellations of the zodiac.

Today, we know that it is not the Sun that is moving through the constellations, but Earth. A little confused? No problem. A quick demonstration will help explain. Take a chair and place it in the middle of the room. The chair will be the Sun and you will be Earth in this demonstration. Stand a few feet away from the chair and look across it to the opposite side of the room. What do you see? A sofa? A TV? Whatever you see will now magically become Leo, the Lion. (It's springtime on Earth!) Next comes summer. Earth (that's you) revolves or walks about a quarter of the way around the "Sun." Look back across the chair. Do you still see Leo, the Lion (a.k.a. your sofa)? Nope. Some other piece of furniture has taken its place. Leo has been usurped by Scorpius (what was once your stereo).

It looks as if the chair (the Sun) has moved in relation to your furniture, even though you (Earth) have been the one doing the moving. Even though Earth has been doing all of the work, people talk about how the Sun moves across the sky through the constellations.

Since Earth follows the same path year after year, the Sun seems to travel through the same constellations. This path is known as the *ecliptic.* Since all of the planets except Pluto orbit the Sun in the same general plane as Earth, all the planets seem to stay within the same constellations as well.

At this point you may be wondering where astrology fits in the picture. Although astronomy and astrology have the same roots, their paths have diverged into two different studies. Astronomy is the scientific study of the stars and planets based on facts, while astrology is a belief or faith, gathering meaning from the mere position of the planets within the constellations. There is no scientific basis for astrology.

WHERE ARE THEY?

Enough about what constellations are. How do you find them? Let's start with a star chart. When you first look at a star chart you may think, "This will be a piece of cake." You will be the neighborhood constellation expert in no time. So after a day of diligent study, you head outside as the Sun goes down.

However, that first look at the real night sky is a sobering experience. Things are a lot bigger in the real world. Not only that, but Mother Nature hasn't drawn any of those nice little lines marking which constellation is which. So your excursion into the night sky lasts all of ten minutes and you return to the house disgusted.

Sound familiar? (This may sound vaguely like the voice of experience, and—guess what—it is.) Because I had been the victim of many star charts in the past, my turning point occurred with the help of a friend. With her help and with some tricks that we picked up along the way, the night sky wasn't nearly as intimidating as it once had been. Read on to learn some of those tricks of the trade.

Whether you use the star chart on the following page, a fancy glow-in-the-dark version, or a high-tech star field on a computer screen, there are a few simple things to do before your first night out that will make stargazing a pleasure.

TRICK #1: START WITH THE BIG DIPPER

I know, I know. It's not a constellation, only a small part of *Ursa Major.* But it's still a great place to start because the Big Dipper can be seen throughout the year, unlike most of the constellations which disappear behind the Sun for periods of time. Also, the seven stars forming the Big Dipper are easy to identify, both in the sky and on the star chart.

Mid January (Approximately 6:00 p.m.)
Mid February (Approximately 7:00 p.m.)

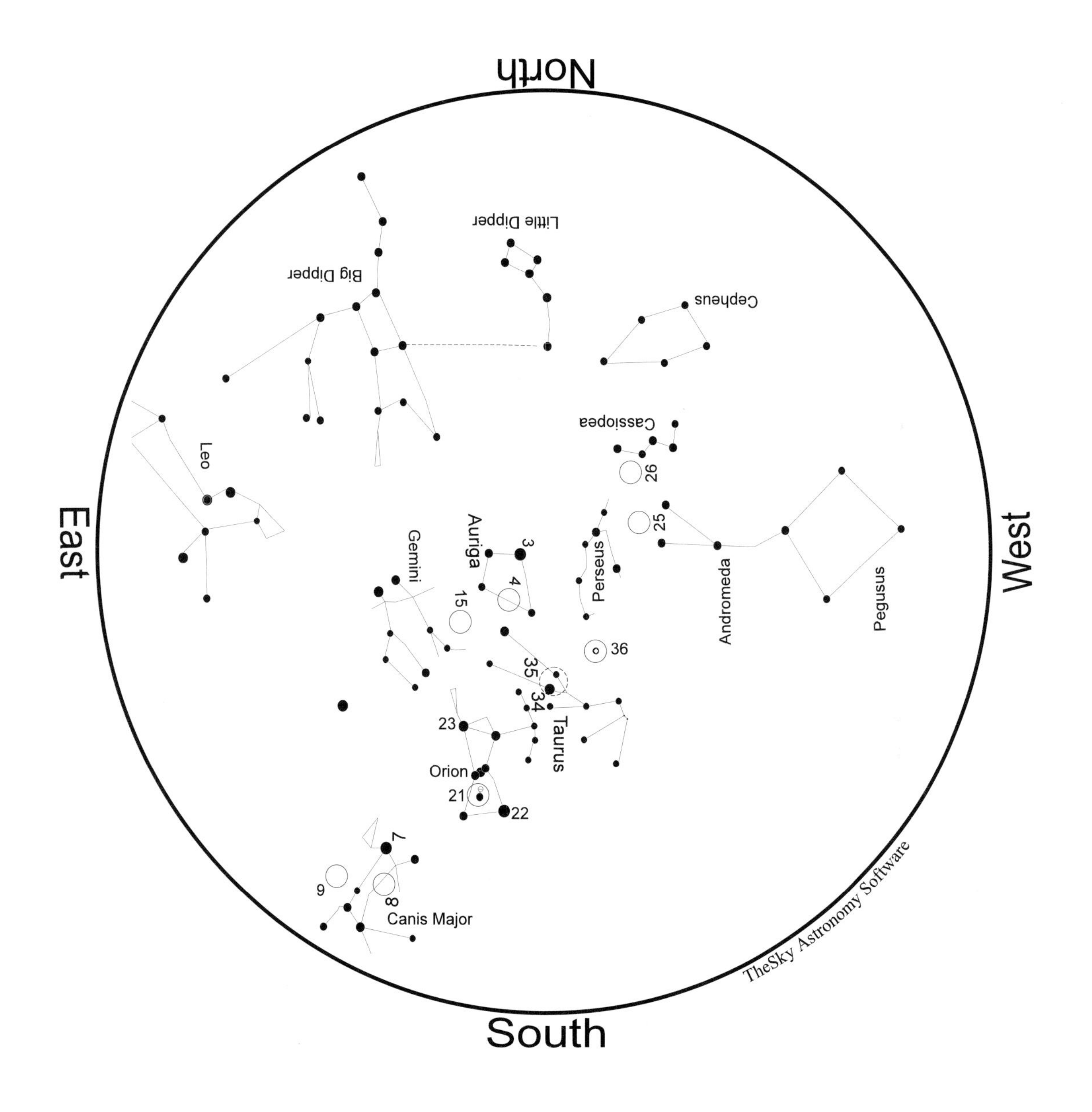

EXAMPLE OF A STAR CHART

Easy identification is the key factor. Any star chart you buy will look different from any other. Stars will be bigger or smaller, farther apart or crammed together, with lines or without. If you can't recognize a group of stars you are already familiar with, how are you supposed to learn new constellations? If you are not familiar with the Big Dipper, don't worry. You will be soon.

Locate *Ursa Major,* the Big Bear, on your star map. Some star maps may have the Big Dipper labeled as such. Most won't. However, they will usually have lines connecting the seven Dipper stars marked within the constellation *Ursa Major.*

Notice in which part of the sky it will appear at the time you intend to observe. Generally, the Big Dipper lies in the northeastern sky as the sun goes down in the spring. By summer, the Dipper has moved until it is high overhead when the sky grows dark. In the fall, the Dipper is hanging upside down in the northwestern sky after sunset. During November, December, and January some people in the South may not see the Dipper for a while after sunset because it is lying so low on the horizon. In December and January, you may have to wait until after 10:00 p.m. before you can make out the Dipper rising in the northeast.

Once you have found the seven stars of the Big Dipper on the chart, study them. How close do they appear to be to each other? Do some stars appear bigger (brighter) than others? Study the stars even if you already know what the Dipper looks like in the night sky. This technique will help you when you move on to other constellations.

Now that you have studied the chart and figured out which part of the sky you should be looking in, it's time to go outside and take a look.

TRICK #2: KNOW WHEN TO LOOK AND WHERE TO LOOK FROM

The obvious time to look is at night. You won't see many constellations in the daytime. There are some nights, however, that are better than others. It all depends on the Moon.

A Full Moon may be beautiful to look at, but it is rather selfish. It not only steals the spotlight away from any star, it *is* the spotlight. When there is a Full Moon in the sky, that is pretty much all you are going to see. So for the few days before and after a Full Moon, study it instead of the constellations.

Now, where should you look from? Your backyard will probably work at first, provided, of course, you can see the sky from there. The more of the sky you can see, the easier it will be to get your bearings. Since you are starting with the Big Dipper, you should have a clear view of the northern sky.

You normally want a dark sky when you do any astronomical observing. But when you first start looking for constellations, a few street lights may actually do you a favor. A few lights will drown out the fainter stars and leave only the brighter ones. Constellations are generally made up of the brightest stars. With the fainter ones out of the way, the only stars you see will be the ones that make up the constellations.

After you become familiar with a few constellations, you may want to move to darker skies where more stars are visible. For the first few times, the most important goal is to find an area where you can see a large portion of the sky.

Trick #3: Think big!

When you study a star chart, you are holding the entire night sky in your hands. This godlike view of the stars tends to make things a bit confusing when you look at the real thing. When you step outside you are once again a tiny mortal with the entire night sky covering you like a huge blanket. Those seven stars of the Dipper that were clumped together in a nice, small, manageable group on your star chart now cover a huge portion of the sky.

After you find the Big Dipper (or when you think you find it, anyway) study the stars as you did the ones on the star chart. How close do they appear to be to each other? Do some stars appear bigger (brighter) than others? Do you notice anything unusual about any of them?

Comparing the Big Dipper stars in the real sky with how they look on the star chart will help you when you move on to other constellations.

Just a note of interest: You might notice something unusual about the second star in the handle of the Dipper. Named Mizar, this star is what astronomers call a visual double star. If you stare at the second star in the handle, you may see a tiny star that appears to be riding piggyback on the brighter one. It is said that the second, fainter star was used to test a person's eyesight in the olden days. If you could see the piggyback star, you had good eyesight.

TRICK #4: USE CATCHY PHRASES TO FIND OTHER CONSTELLATIONS

Follow the arc to Arcturus, then speed on to Spica. Follow the pointer stars to Polaris. Hercules is fighting Draco, the Dragon; and Cygnus, the Swan, is flying along the Milky Way.

Using catchy phrases is just one of several ways to help you remember how to find different constellations. As you expand your knowledge of the night sky, use the constellations you are familiar with to guide you to ones you don't know. Take the Big Dipper, for example. Notice the two stars at the front end of the scoop.

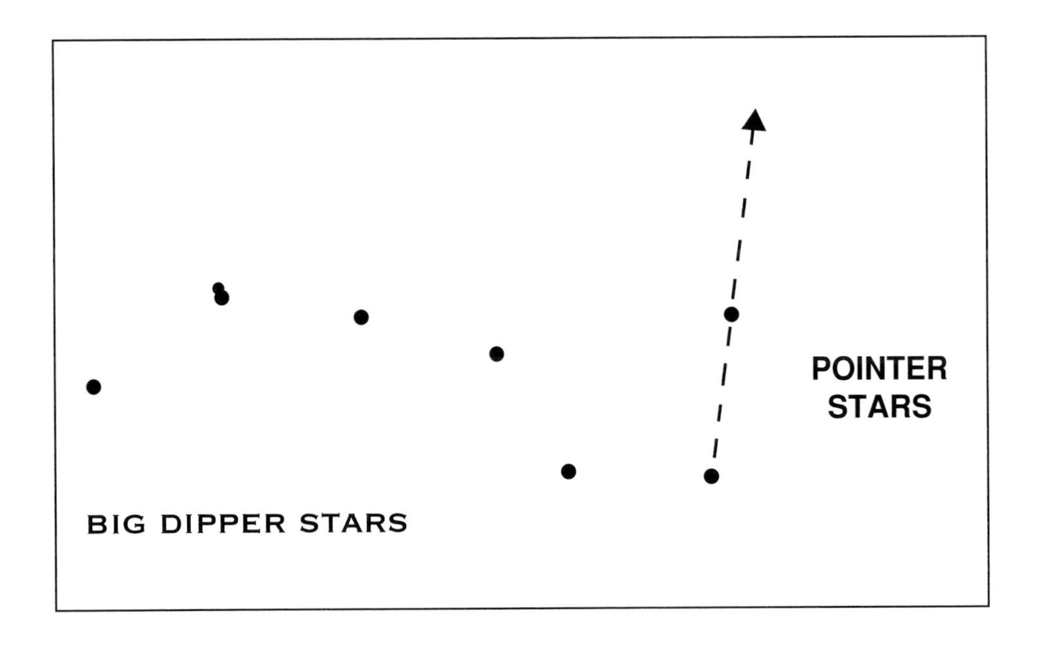

The two stars that form the front end of the Dipper are called the pointer stars. Why, you may ask? Because if you draw a line between the two stars and keep going—heading out away from the Dipper and not underneath the bottom of the Dipper—they will point you to the star Polaris. Big deal, you mutter under your breath. Yes, as a matter of fact, it is a big deal. Polaris is much more commonly known as the North Star. The North Star appears fixed in our skies, pointing north while all of the other stars in the sky rotate around it. The North Star is not the brightest star in the sky, not by a long shot. But the fact that it remains fixed in our skies gives it important status.

So you follow the line drawn by the pointer stars of the Big Dipper, and the first star you come to is Polaris, the North Star. But wait, there's more. Not only have you found the North Star, you've also found another constellation—*Ursa Minor,* the Little Bear, or the Little Dipper. Polaris is the last star in the handle of the Little Dipper. Be sure to check your star map before you go running outside to look.

On your star map, notice that the Little Dipper is a lot different from the Big Dipper. Notice especially the size, or brightness, of the stars. Not only is the constellation smaller, but many of the stars in the Little Dipper are smaller, or fainter, than the stars of the Big Dipper. Fainter stars mean they won't be as easy to see, especially if you live in a city or near a streetlight.

Also, compare the size of the two constellations. Sure, the Little Dipper is smaller, but by how much? Is it half as small as the Big Dipper? Does it look as though the Little Dipper will fit in the scoop of the Big Dipper? Just how far away is the North Star from the pointer stars?

These questions may seem tedious at first, but if you take the time to study your star chart before you go outside, you will be more successful in your quest.

After you follow the pointer stars to Polaris, go to the other end of the Big Dipper (on your star charts first, then in the real night sky). If you draw a line connecting the three stars in the handle, the line won't be straight. Instead, it will form a curve or an arc. Now follow that arc, moving away from the dipper, and you will find that it leads you to the star called Arcturus. With that star you have found another constellation, *Bootes.* If you continue that arc you will run into the star called Spica. Spica is located in *Virgo,* yet another constellation.

The pointer stars to Polaris, the arc to Arcturus, then speed on to Spica, and with that you now know four constellations. Only 84 to go!

Constellation Activities

ACTIVITY 1: A constellation a week

Materials Needed: A star chart; a different constellation every week; a library; a notebook; an hour or so to do some reading.

Description: First, pick a constellation from your star chart and locate it in the sky. Then go to the library and research some of the mythology surrounding that constellation. Since most of the constellations come to us from the ancient Greeks and Romans, there are some wonderful stories that go along with those figures in the sky. Write down your findings in a notebook so you can remember the stories and share them with your friends. There are many books about constellations and Greek mythology. You will find that some constellations have more of a history than others. Some of the more famous are *Orion*; *Hercules* and *Draco*; *Scorpius*; and *Perseus, Pegasus, Andromeda,* and *Cassiopeia.* Share the stories you discover with your friends during your next constellation hunt or your next campout.

ACTIVITY 2: Where did it go?

Materials Needed: A star chart; a watch; a notebook; a pen or pencil; your fist; a clipboard or other hard surface to write on; a flashlight with a red filter to protect your night vision.

Description: Go outside at the beginning of an evening and, using your star chart, locate a couple of constellations. Notice where each is in the sky. One way to do this is to note its location in relation to something on the horizon. Perhaps Orion is just above a tree, or Pegasus is rising above your neighbor's house.

Measure how far the constellation is above that tree or house by using your fist. Spread out your hand and hold it at arm's length with your thumb toward the sky. Now make a fist. Place the bottom of one fist at tree level. Is Orion one fist above the tree, or perhaps two? Record in your notebook how many fists Orion is above that tree or how many fists it takes to reach Pegasus above your neighbor's house. Also record the time of your observations.

Go inside, read a book or watch television for a couple of hours and then repeat the process. See if your constellations are in the same place. I bet they won't be where you left them two hours ago. Record their new positions using the same fist technique.

Did any of the stars within each constellation move? Did the constellations really move? The answer to both questions is no. What did move was

Earth. As Earth rotates on its axis towards sunrise, most of the constellations get left behind.

ACTIVITY 3: Same time, same place?

This activity is much like Activity 2, but instead of going out after a couple of hours to check the constellation of your choice, check it once a week at the same time.

Materials Needed: The same as Activity 2—a star chart; a watch; a notebook; a pen or pencil; your fist; a clipboard or another hard surface to write on; a flashlight.

Description: Choose a night that is not too busy. (Friday night is probably not the best choice for those with an active lifestyle.) Also choose a time during the night that you will be able to get away for a few minutes once a week—for example, every Thursday night after your local news.

Now that you have a time, choose a place. For convenience, try to do this activity in your backyard or some place in your neighborhood where you have a relatively clear view of the night sky. You will want to view the constellations from the same location each week.

With a time and a place, all that is left to do is choose a constellation. With your star chart, go outside on your chosen day and time and see what you can see. Pick a constellation that's easy to identify and, using the fist method described in Activity 2, record its position in your notebook.

As you noticed in Activity 2, Earth's rotation causes the constellations to move across the sky throughout the evening. Observing a constellation once a week at the same time and from the same place will show you what the ancient Babylonians first noted but didn't understand. Earth is slowly changing its place among the stars.

As you observe the constellation at the same time every week, it will appear to move a little higher in the east, or a little lower in the west, depending on where it was when you started. In a month's time, the constellation will have changed positions dramatically. A constellation that started along the eastern horizon, for instance, will be much higher in the eastern sky. A constellation that started low in the west will disappear below the horizon.

Earth's rotation is not the cause of this change—Earth's revolution is. The term revolution means the movement of Earth around the Sun. (Earth's *rotation* or spinning on its axis causes day and night.) Earth's *revolution* around the Sun is rather slow. After all, it takes Earth 365 days to revolve around the Sun one time. The movement of your constellation is a direct result of Earth's movement along its orbit.

ACTIVITY 4: Picturing Orion (and all the others)

Capturing the constellations on film is easy. All you need is a 35mm camera and the ability to take long exposures.

Materials needed: Dark skies; a 35mm camera; cable release; a tripod; 400 ASA speed film or faster; a star chart (to check your positions); a flashlight (to keep you from stumbling around in the dark); paper and pencil.

Description: Besides wide open spaces with lots of sky and little blocking the horizon, you need dark skies, without streetlights or billboard lights. The fast film you will be using will pick up any stray light that is around. This is unfortunate, but the fast film is necessary. The stars you are trying to take pictures of are a lot fainter than most objects you have taken pictures of in the past. But hey, everyone needs a challenge, right?

Once you have found an observing site, set up your tripod a short distance away from your car or any well-traveled path. It will be dark and you don't want to accidentally trip over one of the tripod legs.

Before you attach the camera, take one or two pictures of some bright object. The pictures don't have to be perfect—they can be of your car headlight or an unfocused picture of your flashlight. The purpose of these "bright frame" shots is to help the people who develop the film. If all the exposures on a roll of film are of the night sky, developers have a hard time finding the break between exposures on the negative. There have been several instances where developers who were not exactly sure where to start have cut astronomical images in half.

After you have taken a bright frame, attach the camera to the tripod. Also attach the cable release. You will be taking long exposures, and without a cable release your hand may jiggle the camera. Next, set the f/stop to the smallest setting (the smallest number), and set the shutter speed to "B" for bulb.

Use the star chart to pick a constellation to photograph. Loosen the tripod adjustments so you can swing the camera around to the constellation you've chosen. With the constellation in the field of view of the camera, you are ready to shoot. Open the shutter and start counting. You want an exposure time of 25 seconds. Count to 25 (using the "one Mississippi, two Mississippi" method) then close the shutter and advance the film. In the notebook, record which constellation you photographed. Move the camera to another constellation and repeat the process.

By counting to 25 you allow the film to gather enough light so you can make out the stars, but the exposure time is short enough that you don't pick up any of Earth's movement. For an example of Earth's movement, you may want to take one shot that is a couple of minutes long. When the photo is developed, the stars will appear as tiny streaks on the film rather than pinpoints. These streaks are a direct result of Earth rotating on its axis. Your side of Earth is turning towards the Sun while you take the photograph. In a way, it's like taking a long-exposure picture of a soccer game. The players will be blurry because they don't have time to stand around for a couple of minutes during a game.

When you have your pictures developed, make a note on the envelope that these are astronomy photographs and that the sky is supposed to be black. Some developers may take a little extra time to make sure the contrast is correct.

You can use your photographs to compare what you see in the night sky with what you see on the star charts. You can use them to identify stars within a constellation. You can also use them to impress your friends.

2

The Moon

The Moon is the perfect place to start a journey into the world of observational astronomy. It's close, it's bright, it's fun to watch as it moves across the sky or changes shape, and its surface features are easily visible with the help of binoculars. With credentials like that, how can you not love the Moon?

This chapter is devoted to observing the Moon with just your eyeballs. You will learn why you can sometimes see it during the day, why you can't see it every night, and many other Moon facts you can use to impress your children, parents, and best friends. But before we do that, let's start back at the beginning and discover just where the Moon came from.

Way Back When

For many years, astronomers believed that the Moon was formed at the same time as Earth and has always been Earth's companion. However, that idea has changed recently. Why? It has to do with rocks—Moon rocks to be exact. When scientists studied lunar rocks brought back by the Apollo astronauts, they got some interesting, if rather unexpected, results.

It turns out that the Moon's surface is made of many of the same materials that are found on Earth, but the rocks are not identical. The amount of some of the minerals and elements in Moon rocks is very different from that found in Earth rocks. It's as if part of the minerals come from Earth and part come from somewhere else. Where? Have no fear, astronomers have come up with a possible explanation.

It all started a long time ago. The first character in this drama is a young Earth, with a surface of molten lava and thick clouds of sulfuric acid. Young Earth traveled around in its orbit alone, with no moon to keep it company. The second character in this tale is a rogue, a huge piece of debris left over from the formation of the solar system. The rogue was almost half the size of Earth.

While Earth traveled in its orbit around the Sun, the rogue had no such order. It flew through the solar system, ignoring the ordered paths of the planets. One day, the rogue was a little too careless and smashed into Earth. As you can imagine, neither really enjoyed the experience.

The violent explosion that followed destroyed the rogue. (Earth probably didn't feel too good either!) The explosion flung huge amounts of material out into space. This material (debris from the rogue and Earth) didn't entirely escape Earth, though. Instead, it became trapped in orbit by Earth's gravity. There, the material mixed together to form a ring of debris around Earth. Over the next few thousands of years the debris came together to form one large body. This large body, a mixture of material from both the rogue and Earth, is the Moon we see in our skies today. And no, the rogue piece of space junk was not made of green cheese!

Thus, our Moon was born a long time ago. It has been in Earth's skies ever since, adding to our myths and legends.

Now You See It, Now You Don't

If you observe the Moon for a few nights in a row, you will notice a couple of things almost immediately. It changes shape, and it moves across the sky. These two facts alone have fueled many a legend—from a dragon eating the Moon to a god riding its chariot across the sky to a princess slowly covering and then revealing her face—all in an attempt to explain the mysterious movement of the brightest object in the night sky. As with most things, the truth isn't nearly as exciting or imaginative as the stories.

All the changes you see in the Moon are a direct result of one thing—the Moon's orbit around Earth. It takes 29½ days for the Moon to complete its orbit, so it is never in the same place in the sky any two nights in a row. In fact, it is moving all the time. However, you won't notice this movement if you just glance at the Moon every now and then. You need to observe it for a couple of hours in one night. You will see that in relation to the background stars the Moon moves a distance equal to its own diameter in about an hour.

This explains why the Moon moves every night—but what about those changing shapes? How does the movement explain a banana-shaped Moon or a bright Full Moon?

Actually, the easiest way to explain the phases of the Moon is with a little demonstration. You will need a ball and a dark room with one bright light.

Your head will be Earth, the ball will be the Moon, and the bright light will be the Sun. The bright light should be at eye level. (If you don't have these items handy, a really good imagination might work just as well.)

Stand facing the bright light and hold the "Moon" out at arm's length, directly in front of you. The "Moon" should block your view of the light. Can you see any of the sunlit side of the "Moon" (the side of the ball that is illuminated by the light)? No. The sunlit side of the "Moon" is facing directly away from you, or "Earth," showing you only its dark side. This represents New Moon phase.

Now, still holding the Moon out in front of you, turn 90 degrees, so that your right shoulder is pointed towards the "Sun." The "Moon" has now moved a quarter of the way through its orbit. How much of the sunlit side do you see? (If the room is not very dark, it may be hard to see a clear distinction between the light and dark sides on the ball. Try moving your arm to the right and left a few times, watching carefully to see which side is bright and which is dark.) With the "Moon" a quarter of the way through its orbit, you see half of the sunlit side and half of the dark side. This represents First Quarter phase.

Turn so that your back is facing the light. With your left hand, hold the "Moon" directly in front of you and slightly overhead. It is now halfway around the "Earth." How much of the sunlit side do you see? All of it. When do you see the entire sunlit side of the Moon in the real night sky? During a Full Moon, of course.

For a little Full Moon experiment: Leave the "Moon" in its current position and turn your body so that your right shoulder is pointed towards the "Sun" and your left shoulder is pointed towards the Full "Moon." If you turn your head (Earth) to the right, you see the "Sun." If you turn "Earth" to the left, you see a Full "Moon." In the real night sky, a Full Moon always rises in the east as the Sun sets in the west. Earth is in the middle.

Continue with the demonstration, placing the "Moon" in different positions around "Earth." See if you can find a crescent Moon or a bloated quarter Moon. Since your head represents Earth, your shoulders represent the eastern and western horizons. As you place the "Moon" in different positions, turn your head from shoulder to shoulder to see if the "Sun" and "Moon" are above the horizon at the same time. You will find that sometimes you can see them both at the same time. When the Sun and Moon are above

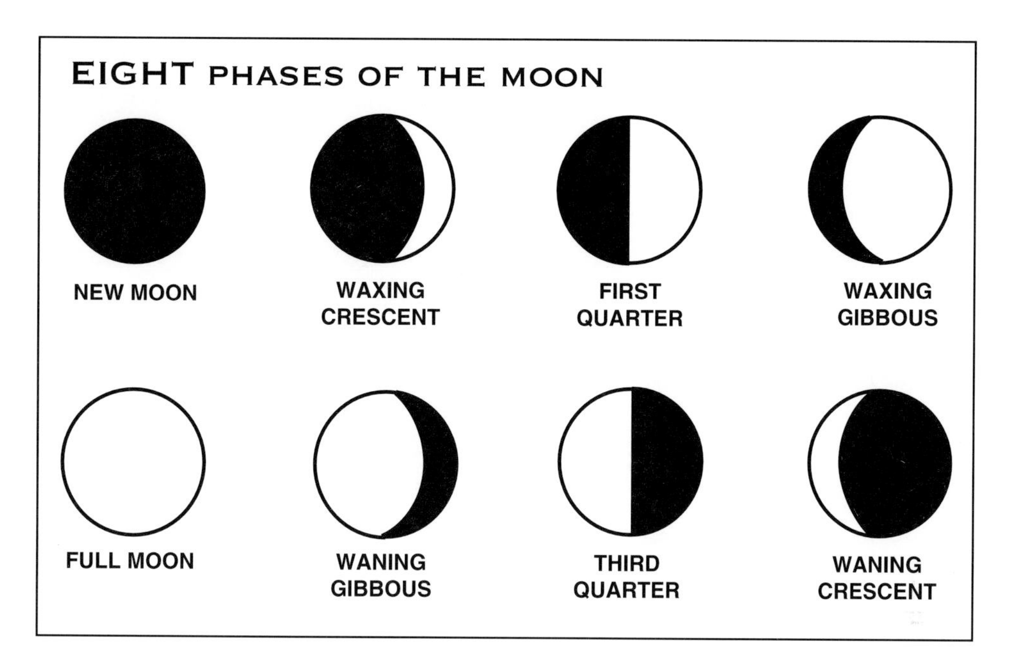

the horizon at the same time, you can see the Moon in the middle of the day.

Phases of the Moon are a little tricky to understand. You can either play around with the demonstration awhile until you have an understanding of the physics of the solar system, or you can simply memorize the following sentence to impress your friends or teachers: As the Moon travels around Earth, we see different amounts of its sunlit side—we identify the different amounts as phases.

The eight different phases the Moon goes through as it travels around Earth are shown above. The term "waxing" means growing—the Moon grows larger in the sky. The term "waning" means the opposite—we see less and less of the Moon.

Lunar Activities

ACTIVITY 1: Full Moon fun

As shown in the earlier demonstration, a Full Moon occurs when Earth lies between the Sun and Moon. From our vantage point on Earth, we see the entire sunlit side of the Moon. A demonstration is great, but it is much more impressive to see the real thing.

Materials Needed: A calendar with Moon phases labeled; a newspaper that gives the time of sunset; a wide-open viewing area with a clear view of the eastern and western horizons.

Description: Once every 2½ years there is a Blue Moon. That is the name astronomers give to the second Full Moon in a month. So, with the exception of that month with the Blue Moon, you will only be able to do this activity once a month. On your calendar, find the day marked Full Moon. On that day, check the newspaper to find out what time the Sun sets below the horizon.

Be at your observing site just before the Sun goes down. Make sure your view of the eastern and western horizon is not blocked by any tall trees or houses. From your vantage point, you will be able to observe the Full Moon rising above the horizon in the east at the same time as the Sun is setting in the west. We are right in the middle.

ACTIVITY 2: Full Moon madness

How many times have you heard someone blame the crazy way people are acting on the Full Moon? Probably quite a few times. Has anyone ever bothered to check if there really was a Full Moon in the sky? Probably not. Conduct your own experiment to find out if there is any truth to the rumor or if the Moon is getting bad press.

Materials Needed: A calendar with Moon phases labeled; a small notebook; pencil; a good pair of ears; a few months to gather data.

Description: Open your calendar to the current month. In your small notebook, record the dates of New Moon, First Quarter, Full Moon and Third Quarter. (You will notice that they are almost a week apart.) Keep this notebook handy, so any time you hear a reference to the Full Moon, you can check the dates and see if there really is one in the sky that night.

Once you have your notebook ready, all you have to do is open your ears and listen. Any time you hear a comment blaming crazy human behavior on the Full Moon, whether you are walking down the street, at work, or at home with the children, record the date in your notebook for a future reference and then compare that date with the day of the real Full Moon.

Do this activity for a few months. The more references you have, the more complete your research will be. Then, after four or five months, add things up. How many references to a Full Moon actually took place on the night of the Full Moon? How many missed the Full Moon by a few days? You may find that, for the most part, the poor Full Moon is getting unfairly accused.

ACTIVITY 3: Does it really move?

In just one night, you can see for yourself that the Moon really is moving around our planet.

Materials Needed: A calendar with Moon phases labeled; a notebook; pencil; two 5-minute sessions two or three hours apart.

Description: Open your calendar to whichever month it happens to be and look for the day marked First Quarter. The phase of the Moon is not really important in this exercise, since all you are interested in is seeing the Moon move. You just need to make sure the Moon will be above the horizon long enough for you to do your observations. For this exercise, choose a night any time between First Quarter and Full Moon.

On the night you have chosen, go outside about an hour after sunset and locate the Moon. Then look around the Moon for any nearby bright stars. If there are no bright stars close to the Moon the first night you observe, you may want to try again the next night. In your notebook, record the time of your observation. Sketch the Moon and where it is in relationship to the bright star. Then go inside and watch your favorite television programs or read a book for a couple of hours.

After your two- or three-hour break, go outside again and look for the Moon. Everything in the sky will have shifted slightly as Earth rotates on its axis. However, when you find the Moon you will notice something different about its position with respect to the star. The distance between the two has changed. The star remains fixed in its constellation, while the Moon is moving along its orbit.

If the next evening is clear, take a peek at the Moon and the star from the night before. You may be surprised just how far the Moon moves in a night.

ACTIVITY 4: Which way did it go?!

For a week or two, with just a few minutes away from the TV each evening, you will see with your own eyes what inspired the Moon myths and legends of our ancestors.

Materials Needed: A calendar with Moon phases labeled; a notebook; a pencil; an open area with a fairly clear view of the southern horizon; a watch; about five minutes an evening for a week or two; a magnet; a refrigerator.

Description: To begin, choose your starting night. Rather than randomly plucking a day out of the air like Uncle Albert's birthday or your parents' anniversary, open your calendar to the current month and look for the day marked New Moon. Actually, the day you really want is a couple of days after New Moon. If that day has already passed, don't worry. You still can't use Uncle Albert's birthday, but you can begin on the day marked First Quarter.

Next, choose your observing area. Look for a place that is easily accessible with a relatively clear view of most of the southern horizon. You are going to be observing for a few minutes every night for a week. It would be inconvenient if your observing site is 10

miles away from your house, so stick with your back yard or somewhere in your neighborhood.

Choose an observing time that is easy to remember—after the news or before the *Star Trek* reruns, for example. The time should also be after dark. When the date, time, and location are set, write them on a piece of paper and, with the magnet, stick it to your refrigerator door as a reminder. (I bet you were wondering where the refrigerator would fit in.)

On the chosen date and time, armed with a notebook and pencil, you are ready to begin. Go outside and look for the Moon. (Hint: If you begin observing just after sunset a couple of days after New Moon, look for a thin crescent Moon in the southwestern sky. If you begin observing on First Quarter, look for a half moon high to the south.) Sketch the Moon's position relative to houses, trees, and other things on the horizon.

As you observe throughout the week, you will see the Moon pass gracefully across the sky, growing larger and brighter.

3

The Planets

Warning: While Venus rising high in the early morning sky can be a stirring sight, planet-watching for the most part is not as dazzling as you may believe. Today we have become spoiled by beautiful closeup pictures of our planetary neighbors. The gorgeous photos you see in magazines or textbooks were taken by spacecraft, either orbiting above Earth's atmosphere or traveling near the planet itself. From your back porch, all you see of the planets are pinpoints of light. If you have a telescope, you can see a little more detail, but that is for another chapter.

Don't be discouraged, though. Even as pinpoints of light, the planets offer their own observing challenges and rewards. One of the biggest challenges is to actually find a planet in a sky filled with twinkling stars. How hard can it be to find a planet? Harder than you think, if you ignore the planet you are standing on, that is. You can forget using most star charts. Unless the star chart is published monthly, chances are the planets won't be listed.

No, publishers of star charts are not playing a joke on you. They have a very good reason for not listing the planets—they can't keep up with them. Our solar system neighbors are busy traveling around the Sun, and they don't have time to slow down and have their picture displayed in a permanent star chart.

So while all the stars have been fixed in familiar constellations for centuries, the planets wander through the sky following their orbits. But don't expect to go outside one night and locate a planet by its rapid movement across the sky. Airplanes and satellites move that fast, but planets don't. Instead, you can notice a planet's movement among the stars from week to week.

Monthly astronomy magazines are a good way to keep track of where the planets are located in any given month. These magazines can at least point you in the right direction. There are also several astronomy computer programs that can help. Once you are aimed in the right direction, another challenge presents itself. How do you tell exactly which pinpoint of light is a planet and which is a star? It would be nice if someone had put a bright red arrow in the sky pointing to each planet. But it must have been that person's night off when the finishing touches were put on the night sky, so we have to find the planets all by ourselves.

General Pointers

Despite the old adage, "Stars twinkle and planets don't," one planet looks pretty much like the next star. Some planets are bright, some are faint, but without a telescope everything looks like a pinpoint of light. Monthly astronomy magazines or astronomy software packages for your computer can give you a detailed account of planetary wanderings, but there are some general tips to keep in mind that will help you on your planetary quest.

Mercury

Mercury is the closest planet to the Sun, so the small planet is always trailing right behind the setting Sun or rising just before the Sun in our morning skies. With its small orbit, Mercury travels around the Sun faster than any other planet. What this means for would-be Mercury watchers is that the planet doesn't stay in one place very long. Once you think you have found Mercury, look quickly—in a few nights it will disappear below the horizon.

The best time to look for the planet Mercury is just after sunset or just before sunrise (for you early birds). The best place to look for the small planet is right along the horizon from the northwest to the southwest in the evening sky and from the northeast to the southeast in the early morning sky.

Venus

Venus, like Mercury, can be found following the Sun in the western skies after sunset or rising just before sunrise in the morning. Venus, however, is much brighter than Mercury. So much brighter, in fact, that it has been called the Evening or Morning Star. There are times when it is the brightest object in our skies. Venus can only be found in the western skies just after the Sun goes down or in the eastern skies before sunrise.

Earth

You're standing on it!

Mars

With its rusty-red surface, Mars appears as a reddish-orange pinpoint of light as it makes its way across the skies. It's not bound to the horizon like Mercury and Venus.

Since Mars is only half the size of Earth and millions of miles away, it's seldom very bright in our skies. To find the red planet, use the monthly astronomy magazines to point you in the right direction. Then observe the object you believe to be Mars for several days. (A few times over a couple of weeks will be enough.) Notice its position in relation to the background stars. If the reddish-orange pinpoint is moving slowly among the stars, you have found Mars.

Jupiter

The king of the planets, Jupiter shines almost as brightly as Venus in our skies. Jupiter, however, can dominate the entire night sky instead of being bound to the horizon like Venus. So if you see a really bright starlike object rising in the east as the Sun goes down in the west, chances are you are observing Jupiter. Again, check in astronomy magazines to make sure you're looking in the right direction.

Saturn

The most distant of all of the visible planets, Saturn moves very slowly among the background stars. While brighter than many stars, it is not the brightest object in the sky. It isn't as noticeable as Jupiter or Venus. Even Saturn's famous ring system is no help when it comes to tracking down the sixth planet, since the rings can only be seen through a telescope.

Uranus and Neptune

These two planets are too far away to be seen with the naked eye. Through a good pair of binoculars or a telescope they appear as bluish-green dots.

Pluto

This very small, distant planet can only be seen as a tiny pinpoint of light through a telescope.

The more you observe, the more familiar you will become with the night sky. As you learn the various constellations you will begin to recognize which stars belong in which part of the sky and which pinpoints of light are merely visitors. After a while you may not need to check a monthly magazine to determine that the bright object rising in the east is Venus.

4

METEOR SHOWERS

Instead of a bar of soap or an umbrella, all you really need to enjoy a meteor shower is a pair of eyes and a little patience.

Meteor (also known as a falling star or shooting star): a small piece of dust or rock floating in outer space that burns up as it enters Earth's atmosphere.

Shower: a deluge of particles; any object (rain, pebbles, etc.) falling in great numbers; a barrage or bombardment.

From the above definitions, a meteor shower should be a deluge of tiny pieces of rocks and dust falling from the heavens, burning up as bright streaks of light in our skies. Sounds exciting and even a little intimidating.

There have been nights, according to historical record, when it did look as if the sky were falling. Unfortunately these spectacular nights don't happen very often, and only Mother Nature knows the exact time and location of the show. And guess what—she's not telling. She does take pity on us poor mortals, though, and occasionally drops hints.

ANNUAL METEOR SHOWERS

Mother Nature's hints come in the form of predictable annual meteor showers. These are the ones you read about in the paper or hear about on television. Astronomers can predict when to look for meteors, but it is hard for them to predict exactly how many you will see. To understand their predicament, let's start out with what causes a meteor shower.

For the most part, outer space is very empty—so empty, in fact, that scientists call it a vacuum. This is not like the vacuum you use to clean a carpet. For carpet cleaning, a vacuum sucks up dirt and debris left behind by friends, family, and pets. When scientists try to create an environment like that found in outer space, they use a vacuum chamber. Basically, they want to clean all the air out of the chamber.

The best vacuum chambers on Earth don't even begin to match the emptiness of outer space. But even the superclean vacuum of outer space has a little debris floating about. (Haven't you ever missed a spot while vacuuming your carpet?) As Earth travels around the Sun, it's constantly running into that debris. When the dusty debris hits Earth's atmosphere, the friction between the fast-moving dust particle and our atmosphere causes the dust or rock to burn up.

On Earth, astronomers predict that three to four meteors can be seen burning through our atmosphere every hour, day and night. That may sound like quite a few, but with a little mathematics it averages out to a meteor every 15 or 20 minutes. Those 15 or 20 minutes between sightings are a lot of empty minutes to wait, especially if it's cold outside. To make things even more challenging, you have to be looking at exactly the right place at exactly the right time. Take it from me, the most useless phrase in astronomy is "Wow! Look at that meteor!" By the time you get the words out and your friend turns around, the meteor is long gone.

When do we get a meteor shower? Every now and then Earth's orbit takes it through a very dusty part of space. These dusty areas are filled with interplanetary litter left behind by those environmentally unconscious vagabonds, the comets. As Earth travels through these dusty areas, we get bombarded with anywhere from 10 to 100 meteors an hour. That produces an average of one meteor every 6 minutes to one every 45 seconds. During those times, a meteor shower is indeed a light show.

Astronomers can predict when Earth will travel through these comet trails. Earth makes that easy by traveling along the same path every year. What astronomers can't predict is how dusty any given comet trail will be. Comets may be interplanetary litterbugs, but they are not very consistent. A comet is like a giant dirty snowball, an uneven mixture of ice and dust. Sometimes a large cloud of dust melts away from the comet, sometimes it doesn't.

The dust particles that melt away from the comet continue to orbit around the Sun in the same orbit as the comet. As time passes, the cloud of leftover dust spreads out, forming a long trail. Every time Earth passes through these dusty trails we get a meteor shower.

Listed below are some of the bigger showers that occur throughout the year. Included in the list are the names and dates of the shower, when the peak of the shower occurs, and about how many meteors you can expect to see per hour.

ANNUAL METEOR SHOWERS

Shower Name	Range	Peak Date	Hourly Rate
Quadrantids	January 1–5	January 3	40
Lyrids	April 20–24	April 22	50
η (Eta) Aquarids	May 3–5	May 4	20
δ (Delta) Aquarids	July 29–31	July 30	20
Perseids	August 10–14	August 12	50
Orionids	October 20–22	October 21	20
Taurids	November 3–5	November 4	15
Leonids	November 15–17	November 16	15
Geminids	December 10–15	December 13	50
Ursids	December 21–23	December 22	15

HOW TO OBSERVE A METEOR SHOWER

Now that you know what a meteor shower is and when the major showers occur, let's talk about what to do with all that wonderful information. The way to achieve maximum enjoyment while observing a meteor shower is to think COMFORT. Unlike most events in astronomy, there isn't a set time to observe a meteor shower. True, the best night is usually during the peak of the shower, but you can observe any time during the evening, or all night if you want (for those truly dedicated). To make your evening of meteor watching enjoyable, plan ahead by taking the following steps:

Step 1. Prepare your observing equipment.

- ❑ lawn chair or blanket to lie on
- ❑ blanket to cover up with (even in the summer, those breezes can get cool)
- ❑ picnic basket with snacks
- ❑ cold or hot drinks (depending on the time of year)
- ❑ insect repellent
- ❑ radio (optional)

It sounds as if you are preparing for a picnic instead of a serious astronomical observing expedition. In a way you are doing both. Because meteor showers can fill the sky, the best way to view them is to simply lie on your back with your eyes to the sky and enjoy. Binoculars and telescopes are not recommended for meteor showers. They only look at a small portion of the sky at a time, so the full effect of the shower is lost.

Step 2. Locate a suitable meteor-watching sight.

For best viewing, you need dark skies. If possible, drive out into the country. If you can't leave the city, find an area in your neighborhood where nearby street lights are blocked from view. The darker the skies, the more meteors you will see.

A Few Extra Tips

Best time of night

During a meteor shower, you begin to see meteors soon after the Sun goes down. However, for those truly dedicated souls, the best time to observe a meteor shower is after midnight and on into the early morning hours. During those early morning hours Earth is rotating into the oncoming meteors and more should be visible. However, if you can't stay up all night don't worry about it. You should still get quite a show.

Best date

Plan your observing excursion as close to the peak of the shower as possible. The peak is the period of time when Earth is in the center of the comet's trail. There should be meteors swarming all around Earth.

Best place to look

Most of the time, meteors appear at random, with no pattern or common point of origin. The first one you see may be over your shoulder, while the next one may look as if it's coming right at you. But sometimes, especially during a meteor shower, it appears that all the meteors are coming from the same general direction. Whatever constellation happens to lies in that general area becomes the namesake for that shower. For example, the Perseid meteor shower is named after the constellation Perseus, which rises in the northeast after sunset in August.

If you can't find the appropriate constellation, don't worry about it. You will do very well by simply looking to the east after sunset. If there is a pattern to the shower, your eyes will pick it up and you will find yourself turning in the right direction.

Just because a meteor shower gets its name from a constellation doesn't mean that a meteor shower (or a lone meteor, for that matter) has anything to do with the stars of the night sky. Though they may look like shooting stars or falling stars, meteors are just pieces of space debris floating in our inner solar system.

Meteor Shower Activities

ACTIVITY 1: Meteor-counting contest

This activity allows you to check the accuracy of the astronomer's meteor per hour predictions. It also provides a little competitive action for you and your friends.

Materials Needed: Yourself and some friends; a piece of paper and a pencil; a hard surface to write on; a small flashlight with a red filter (if you have one, a red filter over a flashlight is easier on your night vision than a bright white light); a watch or a timer; comfort gear listed under Step 1.

Description: First, establish your comfort zone for meteor shower observing, as stated in Step 1. After you have set up the lawn chairs and strategically placed the cooler for easy access, you are ready to start.

With pen and paper in hand, note the time on your watch or set your timer. You will count meteors for intervals of 15 minutes, 30 minutes, or an hour. After noting the time, look up and start observing. Every time you see a meteor, put a mark or tick on the paper.

There are pros and cons for writing down each meteor as you see it. The plus side of recording the meteors as you see them is that you don't have to remember which number you were on. If you don't record each meteor, you may lose count after a while, especially if it is a very good shower. The negative side of recording is that with your eyes to the skies and no light to see where your last tick mark was, your tick marks will end up all over the page before you are done. Think of this as a challenge to your deciphering skills!

If you observe for 15 minutes, multiply the number of meteors you see by 4. The number you end up with will be the number of meteors you would have seen in an hour. (For example, 5 meteors in 15 minutes equals 20 meteors an hour.) If you observe for 30 minutes, multiply the number of meteors you saw by 2. (For example, 15 meteors in 30 minutes equals 30 meteors an hour.) Compare your sightings with the numbers listed under the chart of Annual Meteor Showers.

Now compare your number with the number of meteors your friends saw. Did you each observe the same number? Whoever of you saw the most meteors can sit back and relax and let the others get you a drink from the cooler. You may not want to count meteors all night. Sometimes it is more fun just to kick back, relax, and enjoy the view.

ACTIVITY 2: Photographing meteor showers

You don't have to be an expert photographer to catch a meteor on film. All you really need is a 35mm camera, a tripod, and a little bit of luck.

Materials Needed: 35mm camera with the capability of taking long exposures; cable release; wide-angle lens (or the lens with the widest field of view you have); a tripod; 400 ASA speed film or faster; a flashlight; a lot of patience and a little bit of luck; your standard comfort gear (remember Step 1); the ability to count to 25.

Description: After setting up the lawn chair and cooler, place your tripod on a level spot on the ground a little away from your main encampment. You want it somewhat out of the way so you don't bump into it once you have everything set up.

Load your camera with ASA 400 or faster speed film. (You would think this is an obvious step, but trust me—just check and make sure the camera is loaded.) As I mentioned earlier, take one exposure looking at the flashlight or some other bright area before you leave for your observing run to help the developers determine where the individual frames are located on the film. After you have taken a light frame picture, place the cable release on the camera and set the exposure time at B (for bulb).

Place the camera on the tripod and tighten into place. Now aim the camera at the eastern sky. Decide if you want to have the horizon (trees, etc.) in the shot or if you want a sky shot. If you want trees in the shot, make sure the area you are aiming for doesn't have a road running through it. Nothing can spoil an astronomical photo like a pair of headlights.

Once you have decided what you want in the picture, lock the tripod into place. Focus the camera until the stars are tiny pinpoints of light. Now the patience and luck come into play. Meteors can be temperamental. Sometimes they come in bursts, several at a time. Other times there may be several minutes between each sighting. And just because you aim your camera at one area of the sky doesn't mean the meteors will cooperate and put on their show in that neighborhood. Patience and luck have a lot to do with the success of this mission.

Taking a deep breath, take hold of the cable release. Move your feet away from the tripod so you don't bump into it in the middle of the exposure. Press the cable release, open the shutter, and start counting. Count to 25 using the Mississippi or potato method (one Mississippi, two Mississippi, etc.). While you count, watch the sky where you aimed your shot. With any luck you will see a meteor flash through that portion of the sky. When you reach 25, close the shutter. Wind the film, check your field of view to see if the camera is still pointing where you want it, take a deep breath, and repeat the process. Don't be bashful—use the entire roll. The trick to getting a good astronomical photo is trying many times.

If you don't see a meteor, don't get discouraged. Sometimes you get a nice surprise when you develop the pictures. A meteor will show up as a streak of light among the pinpoint-like stars. If it looks as if the camera jiggled while the shutter was open and you know you didn't bump the camera, it could have been the wind. If your tripod isn't a sturdy breed, a slight wind could disturb the camera. To eliminate this problem, make sure you have a sturdy tripod.

If all your stars show up as even little streaks instead of tiny pinpoints, the problem is not the wind. Your problem is a little bigger than that. In fact, it's really big. It's the Earth messing up your shots, and the only thing you can do about it is count to 20 instead of 25. Earth is turning on its axis. Even though we can't feel it move beneath our feet, a camera is sensitive enough to pick up the movement after a while. Any exposure that is under 25 seconds doesn't pick up the movement. However, if your exposure is longer, you get streaks instead of stars.

Since you are standing still on Earth, it looks as if the stars are moving. On your photo, all the stars will be elongated spots. The longer the exposure, the greater the Earth's movement and the longer the streaks. If you have a couple of shots left and not enough energy to count, try this: Stop down the aperture of your camera to f/8, aim your camera at a area of the sky, open the shutter, and leave it for five or ten minutes. The stars will appear as long streaks. You may even see some colors. Don't get discouraged if you don't succeed the first time—practice makes perfect.

When you are ready to have your pictures developed, take the film to a photography store where you can talk to the person who will process the film. Explain that these are astronomical photos and that the background should be as black as possible. If the film is processed like all other film, the background may not be as dark, but you should still see be able to see the stars.

ACTIVITY 3: Become a meteor shower authority

After a couple of nights out you may find yourself judging the quality of the shower. Friends will look upon you with envy as you talk about the brilliant bolide you saw with a 10-degree tail. (The what?!)

Materials Needed: Yourself (for accurate record keeping); friends (optional, but fun to have around); a predetermined grading scale like the one below; a piece of paper and a pencil; a hard surface to write on; a small flashlight with a red filter (if you have one); a watch; comfort gear listed under Step 1.

A Sample Grading Scale

1 point—extremely faint meteor, just barely visible
2 points—faint, with a long streak of light
3 points—average brightness (about the same brightness as most of the stars in your viewing area) with an average streak of light

4 points—bright fireball with short streak of light (less than 5 degrees*)
5 points—brilliant fireball with very long streak of light (over 10 degrees* in length)
10 points—bolide, a fireball that you actually see hit the ground

Add an extra point if you see a meteor "bounce" or "skip" across the atmosphere (one point for each skip) and one extra point for each green trail you see.

*To determine the length of the meteor's trail, hold your arm out in front of you with your fingers spread out and thumb pointing up. Make a fist and turn it 90 degrees. At arm's length your fist covers approximately 5 degrees of the sky. There are 180 degrees from horizon to horizon. If the meteor trail you saw was longer than two fists, it was over 10 degrees in length.

Description: Establish your comfort zone for meteor shower observing, as described in Step 1. Give yourself a little while to get your bearings. If the shower is poor, you may have to revise your grading scale for faint, fainter, and faintest meteors. After you have a sense of the action, begin judging. Instead of recording a tick mark for every meteor you see, record a number or grade.

As you become more experienced, you may want to record the time of the meteor next to the grade. That is impressive in your records. Unfortunately, the time it takes to look at your watch and write it down is wasted meteor-watching time. To streamline your observing run, read about the radio and tape recorder listed as optional equipment for the next activity. Once you have observed a few meteor showers, you will develop your own grading scale.

ACTIVITY 4: The ultimate meteor-watching expert

When you become familiar with the constellations, it can be to fun to take a star map out on nights of a meteor shower. Every time you see a meteor, record its path through the constellations on your star chart.

Materials Needed: Yourself and some friends; a star map of the night sky—one that you can write on and not cry about later; a pencil; a hard surface on which to write; a small flashlight with a red filter; a watch; comfort gear listed under Step 1.

Optional equipment: A tape recorder with a blank tape; a shortwave radio capable of picking up WWV—a station that broadcasts the time on 5, 10, 15, and 20 MHz. With the tape player recording throughout the observing run, and the radio faithfully announcing the time in the background, you can have an accurate time record of meteor sightings by simply yelling every time you see a meteor. When you play back the tape, you will be able to determine the exact time of each sighting.

Description: Once again, establish your comfort zone. Set your lawn chair so you will be observing the same area of the sky that is on the star map. Prop the flashlight so it shines on the star map but not in your eyes. (This may be a little tricky.) If you have a tape recorder and radio, set them up before you get the flashlight strategically placed. Make sure they all have fresh batteries, or it may be a short night! Once everything is in place, begin your observing run. Here's hoping you have lots of meteors to record!

5

Eclipses

For us Earthlings, there are two types of astronomical eclipses: solar eclipses and lunar eclipses. During one, Earth's shadow is the thief stealing the limelight. During the other, the Moon is the culprit. Whichever one is doing the dirty work, there is nothing we can do about it but kick back, relax, and enjoy the view.

A Solar Eclipse

Solar eclipses have always had a mysterious aura. In ancient times, instead of enjoying the view, people were filled with a sense of awe or fear of the unknown. After all, would you be calm if you thought the Sun was burning out? Some thought magic and sorcery made the Sun go dark. Others believed a large monster was eating the Sun, or that the gods were punishing them. As in most cases, however, the truth doesn't sound nearly as exciting. During a solar eclipse, what you are seeing is the Moon passing between the Sun and Earth.

Normally the Moon travels around Earth minding its own business. You don't notice anything unusual. But every now and then it passes between Earth and the Sun and blocks our view.

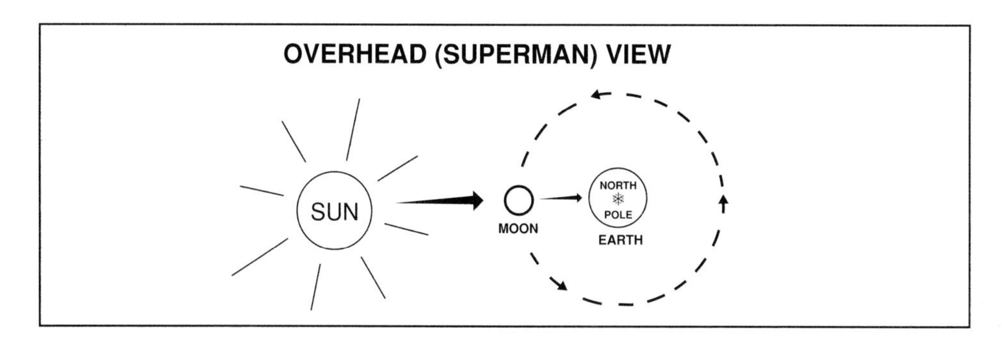

If you were Superman and could fly above the solar system, you would see the Moon circling Earth once every 29½ days. From your point of view it would look as though the Moon blocked Earth's view of the Sun every time it traveled around Earth.

As you may have guessed, this is not the case. After all, you would notice if something other than clouds blocked the Sun for a couple of hours every month. Since you are playing Superman and can fly anywhere, now change your point of view. Instead of flying straight up out of our solar system, fly off to one side. From this angle, it is easy to understand why we don't have a solar eclipse every month.

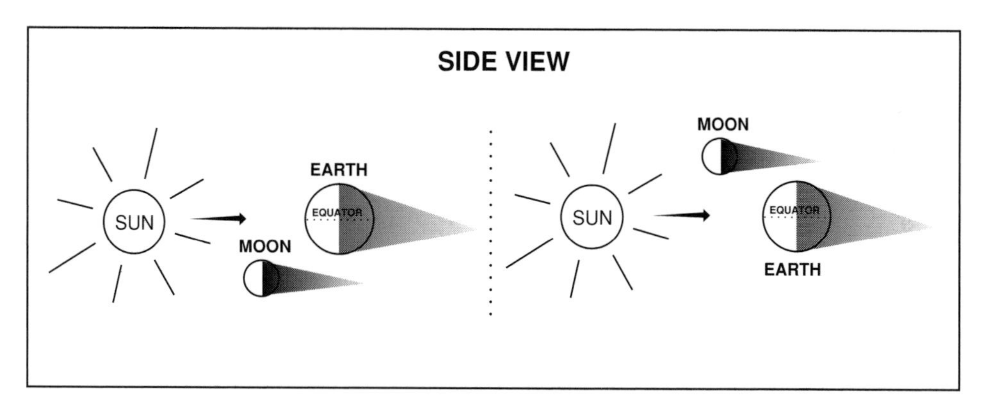

As you can see, the Moon's orbit around Earth is tilted relative to Earth's orbit around the Sun. So sometimes the Moon is below Earth, sometimes above. Those times when it does pass between the Sun and Earth are the times you read about in the newspaper—the solar eclipses.

Types of Solar Eclipse

A solar eclipse is mysterious because you can't see it coming. One minute the Sun is shining and the next a part of it is covered up. The approaching Moon is invisible until it takes a bite out of the Sun. Before there were newspapers and televisions to announce an upcoming eclipse, people were taken by surprise and often startled by these events. (That's how the Connecticut Yankee was able to fool King Arthur's court in Mark Twain's book *A Connecticut Yankee in King Arthur's Court.*) Astronomers, or anyone who tracks the Moon's path across the sky over a period of time, can predict the Moon's orbit and determine when a solar eclipse will occur. Today these predictions are spread across the headlines and a solar eclipse can be enjoyed by all.

The above definition of a solar eclipse—the passing of the Moon in front of the Sun—is a very general description of what happens. If you want to get technical, there are more specific ways to classify what the Moon does to the Sun. The Moon doesn't always cover the entire Sun. Sometimes the Moon appears to take a small bite out of it. Other times, the Moon covers the center of the Sun and a ring of sunlight is visible surrounding the Moon.

A total solar eclipse (the most spectacular). There are no clouds in sight, yet the sky darkens. Chickens begin to roost. Stars come out in the middle of the day. A total solar eclipse is a very impressive sight. For a short time, the Moon completely covers the disk of the Sun. No part of the Sun's disk can be seen from Earth.

Though it may seem longer, a total solar eclipse lasts for about two hours. This does not mean the Sun is completely covered during that whole time. *Totality*—the period of time the Sun is completely blocked—is only a very short time, anywhere from a few seconds to a few minutes. So where do the two hours come from? Well, the Moon doesn't just jump in front of the Sun and jump away from it. It takes about an hour to move in front of the Sun and another hour to move away.

An annular eclipse (almost, but not quite). The only difference between a total eclipse and an annular eclipse is distance. First, an earthly example: Imagine you're in a movie theater. If the person in the row in front of you moves his head he can block your entire view. But if the person three rows in front of you moves his head, he won't block much of your view at all. Why? The guy in the third row in front of you is farther away than the first guy. The farther away he is, the smaller he appears. The smaller he appears, the less of the screen he will block.

Now, back to an eclipse. The Moon not only has an inclined orbit, but it also changes its distance from Earth. On average, the Moon is about 238,800 miles (384,400 kilometers) from Earth. Sometimes it is closer, sometimes it is farther away. When the Moon is at its farthest distance from Earth, it appears slightly smaller in our skies. Normally, it is hard to notice this slight change in size.

The one time you can notice this change is during an annular eclipse. The Moon passes directly in front of the Sun just as it does during a total eclipse, except there is no totality. The Moon, now farther from Earth and smaller in our skies, isn't large enough to cover the entire disk of the Sun. Instead you see a fiery donut. The Moon covers the center of the Sun and a ring of sunlight surrounds it. Even though the Moon covers most of the Sun during an annular eclipse, the sky doesn't get as dark as it does during a total eclipse.

A partial solar eclipse. How many times have you been in the middle of a movie when someone in your row needs to go to the bathroom? If the person is polite, she will try to duck as she passes in front of you. Usually, she is only partially successful. By ducking, she manages to block only your view of the bottom of the screen. You can still see the top of the screen. Your view has been partially eclipsed.

Sometimes the Moon passes across just a portion of the Sun's disk, partially eclipsing the Sun. When this happens, it appears that something is taking a bite out of the Sun. Sometimes it is a big bite, sometimes it is a small bite.

Partial eclipses are rated by the percentage of the Sun's disk that will be covered (20 percent coverage, 60 percent coverage, etc.). The more of the disk to be covered, the longer it will take and the darker the sky will get. While not as impressive as a total or annular eclipse, a partial eclipse is still fun to observe.

Upcoming Solar Eclipses

In order to see a solar eclipse, you have to be at the right place on Earth. In fact, a joke among astronomers says that most total eclipses can only be seen from deep in a desert, a remote high mountain range, or the middle of an ocean. Though this isn't always the case, it seems to happen quite a bit. For a list of future solar eclipses and where you'll have to go to see them, see *Chasing the Shadow,* by Joel Harris and Richard Talcott (Kalmbach Publishing Co., 21027 Crossroads Circle, P. O. Box 1612, Waukesha, WI 53187; phone 414-796-8776).

Solar Eclipse Activities

WARNING: Before you try to observe a solar eclipse on your own, there are some things to keep in mind. If you stare at the Sun for any length of time you can damage your eyes. Even during an eclipse when the Sun is partially covered, you have to be careful. You only have to burn your eyes once to be blind. This may sound alarming, and in a way it is supposed to. The Sun is dangerous and your eyesight is something you want to protect. However, there are some simple things you can do that are easy and allow for safe viewing. The following activities are designed for enjoyable and safe eclipse viewing.

ACTIVITY 1: Creating your own total eclipse

To understand a total solar eclipse, you can create your own.

Materials Needed: A small ball; your head; a bright flashlight. (If you don't have a ball, your hand will work just as well.)

Description: The flashlight will represent the Sun. Your head will represent Earth and the ball (or your hand) will be the Moon.

Turn the flashlight on and set it across the room. If you have another person, you can designate someone to hold the "Sun."

Stand facing the flashlight. Hold the ball in your right hand, off to the side. While you are looking at the "Sun," slowly move the ball in front of you. If you happen to have the "Moon" lined up with the "Sun," the ball will eclipse the flashlight as it moves in front of your face. The light will disappear.

Try raising and lowering the "Moon" as it orbits your Earth. See how exact the alignment has to be for an eclipse to occur. Also, try bringing the "Moon" closer to Earth and see if you notice any changes. Hold the "Moon" as far away as possible and see what happens. You may see an annular eclipse. What do you have to do to have a partial eclipse?

ACTIVITY 2: Observing a solar eclipse with a projection box

The next time you hear about a solar eclipse, you can make your own solar observatory. Remember, it is dangerous to look directly at the Sun. Follow the directions carefully. In this activity, you won't even be facing in the same direction as the Sun. In fact, you will have your back turned to the Sun most of the time.

Materials Needed: A rectangular cardboard box (the larger the better); a razor blade or a pair of scissors; a 3-inch square piece of aluminum foil; a straight pin; a sheet of white paper; tape; pencil; ruler.

Description: Set the box on end with the open side towards you and the bottom of the box away from you. (See diagram.) With the pencil, draw a 2-inch square on the upper left-hand corner of the box end, about 1½ inches from the side and 1½ inches from the bottom of the box.

With the razor blade or scissors, carefully cut out the square you drew. On the inside of the box, tape the piece of aluminum foil over the hole. Carefully poke a small hole in the center of the aluminum foil with the straight pin.

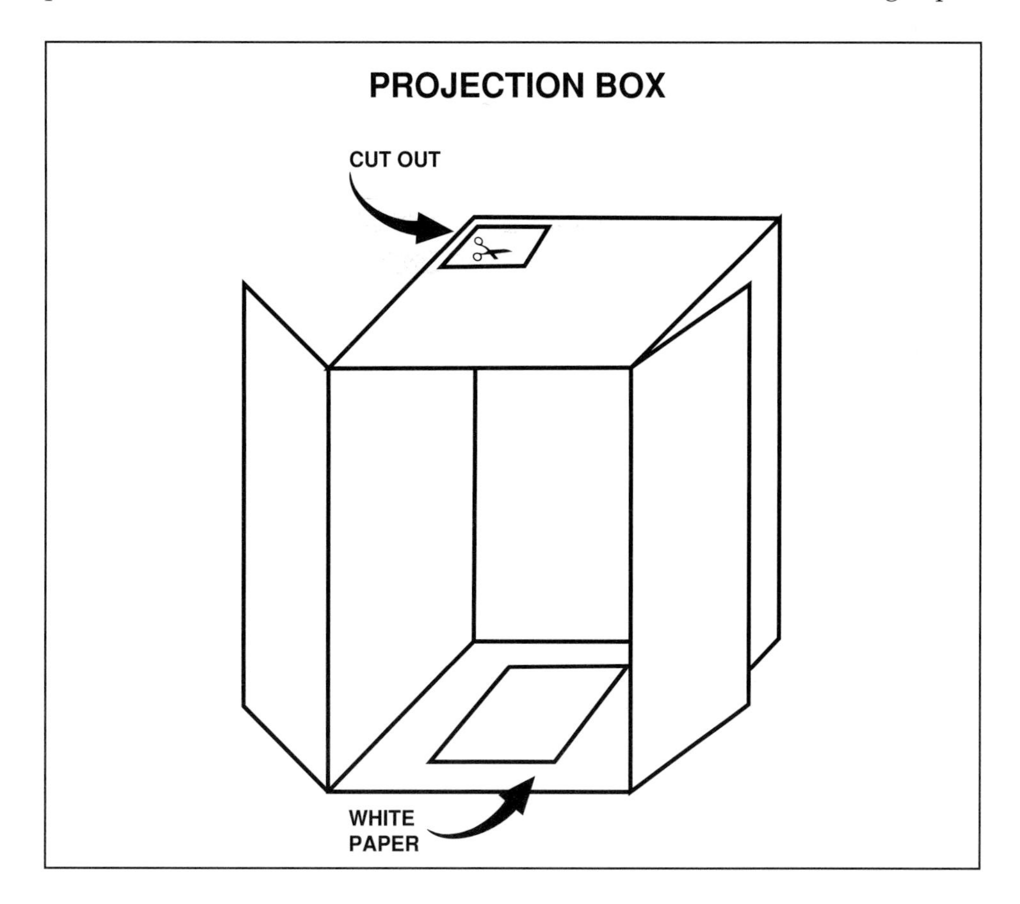

Tape the white piece of paper inside the box on the side opposite the hole. You now have a solar observatory.

How to Use Your Projection Box: You don't need a solar eclipse to use a projection box. In fact, you may want to practice with the box before the eclipse to see how it works.

Take your projection box outside on a clear day. (If clouds are blocking the Sun, forget it! No matter how good your box is, you won't see anything.) Turn your back to the Sun. Hold the box so the pinhole is pointed toward the Sun and the open side of the box is toward you. You should be able to see the white paper without blocking the light coming from the pinhole. It may be easier to stick your head inside the box, off to one side so you don't block the pinhole.

Aim the end of the box in the general direction of the Sun and look at the shadow of the box on the ground. Slowly move the box back and forth until its shadow as small as possible. This will help you align the pinhole with the Sun. When the pinhole on the box is aligned with the Sun you will see a round spot of light. This spot is the Sun's disk. The image will be rather small—the size of the Sun's disk depends on the size of your cardboard box, or more specifically, the distance between the pinhole and the white paper. If you use a small box, the image will be small.

When the Moon is not taking a bite out of the Sun, it is hard to believe what you are seeing with the projection box. That spot is actually the Sun? During an eclipse, it is much easier to believe. You can see exactly when the Moon starts to move in front of the Sun. A portion of that simple round spot will disappear.

A projection box is easy to make and completely safe to use. Although the image of the Sun will be small, during an eclipse a small image will still be impressive.

ACTIVITY 3: Observing under a shady tree

If you live in an area with many trees, you have an easy way of observing an eclipse without doing any work. All you need to do is walk outside and look at the ground beneath the tree. If you are lucky, you will be able to see several images of the Sun, each with a bite taken out of it.

Yeah, right. How many times have you seen the ground covered with suns? Probably more times than you think. The next time you go for a walk on a sunny day, look at the ground under a tree. (Leafy trees work better than trees with pine needles.) What you are looking for is that "dappled" appearance of part sun–part shade. That dappled appearance occurs when sunlight passes through several layers of leaves. If you look closely, you will notice that some of the spots of light are round. If there is a large area of light, you may be able to pick out several overlapping round spots.

These round spots are images of the Sun's disk. Tiny spaces between the overlapping leaves act as pinholes, projecting an image of the Sun. Normally, you don't notice the Sun's images scattered around on the ground. However, during an eclipse the images stand out because the individual round spots of light have bites taken out of them.

ACTIVITY 4: Recording the eclipse on paper, while sitting under a tree

Materials Needed: Several sheets of white paper; book or other hard surface to draw on; pencil; watch; lawn chair; picnic basket (optional).

Description: Before an eclipse, scout around under a few trees. You are looking for a nice dappled shadow as described in the previous activity. Once you have your spot picked out, you are ready to go.

Before the eclipse begins, find the best image of the Sun's disk on the ground and place the drawing surface and a sheet of paper so that the image is projected onto the paper. Just before the eclipse begins, trace the image of the Sun onto the paper and record the time on the bottom of the page. Place a new sheet of paper on the drawing surface.

As soon as you notice a change in the Sun's appearance trace a new image and record the time of the drawing. Continue this process every few minutes (two- to five-minute intervals). Be sure to record the time at the bottom of each drawing.

When the eclipse is complete you will have a picture log of a solar eclipse. You can then bind the drawings in a notebook. Flipping through the pages provides a speeded-up version of the eclipse. (This also makes a good science project!)

A Lunar Eclipse

It is the middle of the night. A brilliant Full Moon shines down on the countryside. You decide to take a quiet stroll through the romantic night. All is calm, you think. A casual glance towards the Moon brings you to a halt. You look again. The second view confirms the first. Something is eating the Moon! You stand helpless, watching as the Moon slowly disappears. Will it come back or will the sky be dark forever? A scene from a new science fiction thriller, perhaps? It could be. Movie or not, this scene has been played out many times in real life in the form of a lunar eclipse.

Observing a lunar eclipse is not nearly as complicated as observing a solar eclipse. After all, how difficult can it be if you can see an lunar eclipse while taking an evening stroll? All that is required to view a lunar eclipse is: 1) a clear night and 2) the Moon visible above your horizon. A few other items (radio, lawn chair, cooler, insect repellent, etc.) will make the lunar eclipse a little more comfortable, but are not absolutely necessary.

What really happens during a lunar eclipse? Quite simply, the Moon moves through Earth's shadow. (The term "lunar" comes from Luna—the Latin name for our Moon.) True, that description isn't as spectacular as something eating the Moon, but that's the way things go!

Like Peter Pan, and anything else with a light shining on it, Earth has a shadow. (Unlike Peter Pan, Earth has never lost its shadow!) As the Sun

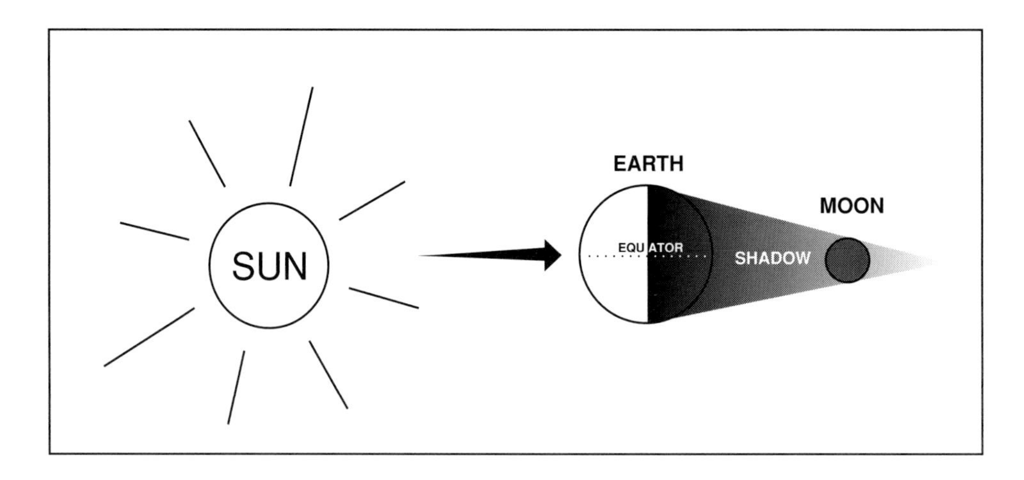

shines on Earth, which is over 7,650 miles (12,756 kilometers) across, it casts a rather large shadow. Most of the time the Moon's orbit takes it above or below the shadow. Every now and then, though, the Moon has no choice but to pass through the darkness. When this happens, we see a lunar eclipse.

Types of Lunar Eclipse

Total lunar eclipse. During a total lunar eclipse, the entire Moon passes into Earth's shadow. Without any sunlight shining directly on its surface, the Moon darkens considerably. You will still be able to see it, though. Light refracted by Earth's atmosphere allows us to see a much darker Moon for the short time it's out of the Sun's spotlight.

The length of time the Moon is within Earth's shadow varies. An average time is 2½ to 3 hours. The Moon is not completely covered for all of that time. It takes about an hour for it to move into the shadow and another hour to move out. During totality, when the Moon is completely within the shadow, it may seem frozen in time, not moving. It only appears frozen. The Moon continues along its orbit, passing through the deepest part of Earth's shadow.

How long it remains in Earth's shadow depends on which portion of the shadow it passes through. After all, Earth is a fairly good-sized planet. It has a large shadow. If the Moon passes through the center, totality can last for

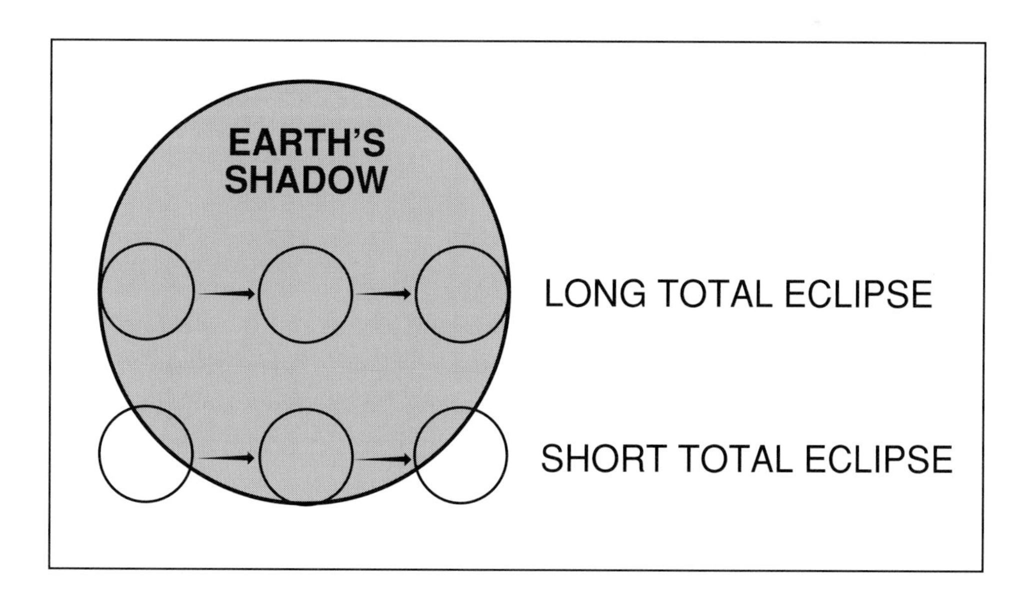

almost an hour. On the other hand, if it passes just inside the shadow, totality will last for only a few minutes.

Partial lunar eclipse. Just as the name indicates, the Moon partially enters Earth's shadow during a partial lunar eclipse. Earth's shadow appears to take a bite out of the Moon. That "bite" moves across the bottom or the top of the Moon, depending on whether the Moon passes slightly above or below Earth's shadow.

Upcoming Lunar Eclipses

Below is a list of upcoming lunar eclipses, including dates, times, and whether it is a total or partial eclipse. With no clouds and a good, open view of the sky, you can kick back, relax, and enjoy Mother Nature's show. The list was compiled from the *Fifty Year Canon of Lunar Eclipses: 1986–2035,* by Fred Espenak.

Date	Type of Eclipse	Time of Mid-Eclipse*
September 27, 1996	Total	2:54
March 24, 1996	Partial	4:39
September 16, 1997	Total	18:47
July 28, 1999	Partial	11:34
January 21, 2000	Total	4:44
July 16, 2000	Total	13:56
January 9, 2001	Total	20:21
July 5, 2001	Partial	14:55

* The time for the Eclipse is given in Greenwich mean time. To find eastern standard time, subtract 5 hours. For central standard time, subtract 6 hours. For mountain standard time, subtract 7 hours. For Pacific standard time, subtract 8 hours. If the eclipse takes place during daylight savings time, subtract one less hour from each time zone. For example, to find Pacific daylight time, subtract 7 hours.

Lunar Eclipse Activities

ACTIVITY 1: How to observe a lunar eclipse

The key to enjoying a lunar eclipse is to properly prepare for an evening outside. Even though you won't need any special observing equipment, you will want to have some comfort supplies. Remember, a lunar eclipse can last for hours.

Materials Needed: A lawn chair or blanket to sit on; insect repellent; ice chest with snacks and drinks or a thermos with hot chocolate or coffee (depending on how cold it is outside); warm clothes (even in the summer, a slight breeze can feel cool after an hour of sitting); flashlight; friends (the more the merrier).

Note: You don't need a telescope or pair of binoculars to observe a lunar eclipse. In fact, using a telescope can limit your view to only a small part of the moon at a time. During a lunar eclipse, it's best to be able to see the entire Moon.

Description: From the materials listed above it sounds as if you're preparing for a picnic. Actually, that's a good way to think about a lunar eclipse. There will be plenty of time to relax and talk with friends. If you begin the evening armed with all of the comforts listed above, the eclipse will prove to be an enjoyable event, provided the clouds stay away. But forget the insect repellent, for instance, and you will be scratching and itching so much that a two-hour eclipse will seem to last forever.

ACTIVITY 2: How to determine that Earth is really round

By playing with shadows, you will see that you can determine an object's shape just by seeing its shadow. Using this method, you can determine Earth's shape by viewing a lunar eclipse.

Materials Needed: A bright light source; a ball; a square block; any other odd-shaped object; a blank wall; round paper cutout representing the Moon (you can draw craters on the Moon if you like); tape; one, two, three, or more people.

Description: If you have ever made shadow figures with your hands, you can do this activity. Set up the light source in the center of a room. (A lamp without a light shade will work very well.) Stand between the lamp and the wall, facing the wall. You should see your shadow on the wall.

If there is more than one person, have everyone sit facing the wall with their backs to the light source. Designate one person to hold up the mystery objects. One at a time, hold up each object. The more unusual the object, the better. Let everyone try to guess what the object is by observing its shadow.

What does this have to do with a lunar eclipse and the shape of Earth? Glad you asked. After you have experimented with several different objects, tape your paper Moon to the wall and repeat the above exercise. This time, the objects will be casting their shadows on the paper Moon.

Compare the shadows cast on the paper Moon with the drawing of a partial lunar eclipse. Which object matches the curved shadow seen during an eclipse? You will find that the round ball best matches the shadow cast upon the Moon by Earth. Thus, by observing Earth's shadow, you can conclude that Earth is round.

6

Comets

The year is 2062. In the news, on televisions and newspapers everywhere, the headlines scream out at you as you cruise by in your wheelchair:

HALLEY'S COMET RETURNS
BIGGER AND BRIGHTER THAN EVER

Just how are you going to view this ageless wonder? Where are you supposed to look? Do you need special equipment? There seems to be a lot of things to worry about, but don't get excited. (At your now rather advanced age, the strain wouldn't be good for your heart.) Just take a deep breath, relax, and read on. As you'll see, Halley's Comet (and any other comet, for that matter) will give plenty of notice before it puts on its show. And for the brighter comets you don't really need anything but your eyeballs to enjoy the view.

Just What Exactly Is a Comet?

Before we talk about how to observe a comet, it might be a good idea to know exactly what it is you will be trying to see. In the 1950s, a Harvard astronomer named Fred L. Whipple came up with the perfect, although rather drab, description of a comet. He called it a dirty snowball (a very big, very dirty snowball, to be precise). But don't worry. When a comet gets close enough to Earth to make the evening news, you are going to be seeing much more than a tiny blob of dirty snow.

With any luck, you will see a beautiful tail streaming away from the bright, fuzzy head of the comet. That fuzzy head is called the *coma.* The coma is made up of ice and dust that has melted away from the dirty snowball (what astronomers call the *nucleus*). The coma surrounds the snowball and blocks our view.

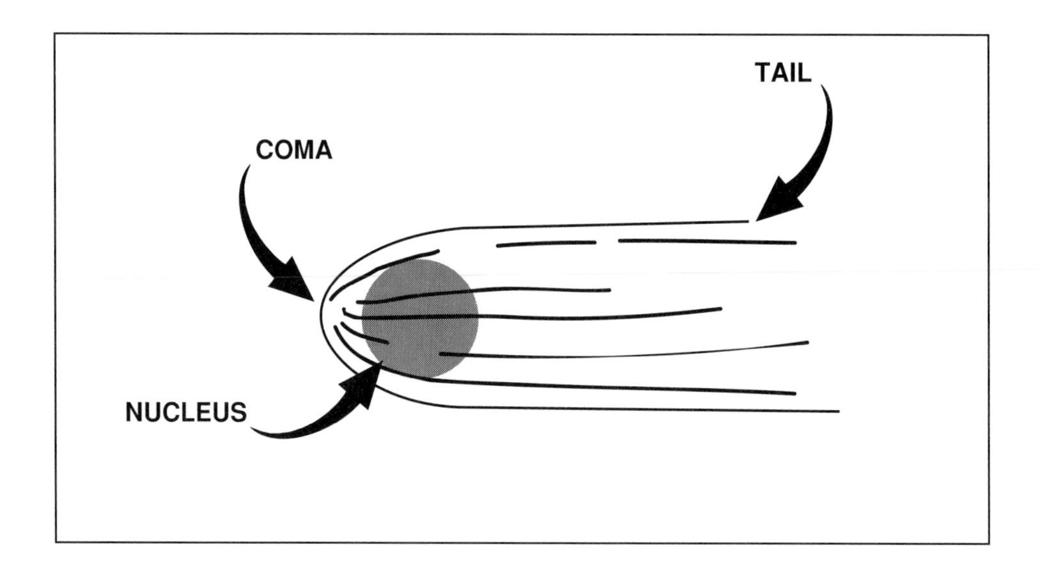

Streaming away from the *coma* is the comet's tail. Strangely enough, you can't judge a comet by its tail. In other words, you can't always tell which way it's moving by the direction of its tail because the tail can be either behind the coma or in front of it.

At first this may seem a little confusing. A comet is traveling through space at high speeds. One would think the ice and dust melting away from the comet would get left behind as the comet speeds through its orbit. This sounds logical, but it's not the reason a comet has a tail. So what's the secret of a comet's tail? The Sun.

The Sun gives off tremendous amounts of energy every second. We experience this energy all around us in the form of heat and light. Energy that we can't see is also released, and some of this forms what scientists call the *solar wind*—charged particles streaming away from the Sun.

Everything in orbit around the Sun (including Earth) passes through the solar wind. We see its effects on Earth in the form of the aurora borealis and aurora australis, the northern and southern lights. But what, you may ask, does this have to do with a comet's tail? It has everything to do with it.

As a comet approaches the Sun it has to travel against the solar wind. It is as if the comet is traveling upstream. The solar wind—blowing away from the Sun—catches the particles of ice and dust within the coma and pushes them away. Since these particles are much smaller and lighter than the nucleus, the solar wind has more of an effect on them than on the large snowball.

You probably have experienced something like this on Earth. Have you ever tried walking against the wind while holding a stack of papers? No matter how tightly you hold them, the wind always seems to grab one or two pieces. The next minute, you are running back the way you just came, chasing your runaway papers.

A comet doesn't chase the material that is caught up by the solar wind. Instead, it goes along its merry way, leaving a dusty trail behind. After it has passed around the Sun and begins to make its way back to the outer reaches of the solar system, the comet begins to travel with the wind—or downstream. Now the tail of the comet is ahead of the coma.

Using the same example as before, imagine you are now walking in the same direction as the wind is blowing. Walking is easier with the wind at your back. If you drop a piece of paper, the wind will still carry it away from you. But this time, the paper will be traveling in the same direction as you. It may even get to your destination before you do!

To make it short and sweet: If the comet is approaching the Sun, the tail will be streaming behind it. If the comet is moving away from the Sun, its tail will be pushed in front of the coma. In the night skies of Earth, with the naked eye, it's difficult to tell the difference.

When, Where and How to Observe Comets

Now that you know what makes up a comet, let's talk about how to observe one. The key is patience and the ability to expect the unexpected. If astronomers know the orbit of a comet they can predict with great accuracy when and where it will appear in our skies. What they can't predict with much success is how bright the comet will appear. Comets, like everything else, have their good days and their bad days, especially when it comes to looks. All astronomers can do is wait and watch with the rest of us.

Why the unpredictable appearance? Does Mother Nature like to keep us guessing? Could be. Astronomers have a slightly more practical explanation, though. They have determined that the brightness of a comet depends on two things: how close it gets to Earth and how much ice it contains in its nucleus.

Since astronomers do a pretty good job at calculating comet orbits, the problem seems to be with the second part—how much ice. Comets, as I mentioned earlier, are dirty snowballs—a mixture of ice and dust. What sort of a mixture depends on the individual comet. If it contains a large amount of ice it will be brighter in our skies than a comet with a large amount of dust.

Why? Ice particles reflect sunlight much better than dust. If you have ever gone skiing without sunglasses you are probably well aware of that fact. The sunlight reflecting off the snow can give you a blinding headache, something you don't experience standing in a dusty field.

When it comes to comets, astronomers can't tell if the mixture has more ice or dust. Sometimes a large amount of ice is released at once, which causes the comet to increase in brightness dramatically. Other times the opposite happens. A large patch of dust may be released, causing the comet's brightness to fade. Astronomers still try to predict what will happen, but it's pretty much up to Mother Nature. When you read about a bright comet in the newspaper, keep in mind the following tips:

Observing Tip 1: Don't expect to see the comet blazing across the sky at high speeds.

True, a comet travels at high speeds around the Sun. But it is so far away from us that we don't get a sense of that speed. Here's an earthly example: Have you ever watched a jet flying high overhead? From your vantage point on the ground, the jet appears to travel slowly across the sky, when it may be traveling at speeds of well over 500 miles per hour (800 kilometers per hour).

A comet is much farther away from us than a jet, so its movement across the sky is much slower. After observing a comet in the sky for several hours, you may detect some small movement if you have really good eyesight. Compare the comet's location with the background stars. If you don't notice any movement over the course of a few hours you will notice some movement from night to night. If you miss it one night, don't worry. It will be there the next night.

Observing Tip 2: Get away from city lights, if possible.

As with many things in astronomy, the darker the sky the better.

Observing Tip 3: The best comet viewing is done with the naked eye or a pair of binoculars.

If a comet is bright enough to make the evening news, chances are it will be much too large to see with a telescope. A telescope is designed to observe a very small portion of space. For example, have you ever been looking through a telescope or a pair of binoculars and had someone walk right in front of you? They are so close, they not only block your view, they overwhelm the binoculars or telescope. You can't begin to tell who the culprit is until you move away from the telescope and see them without all the magnification.

A comet acts as the person blocking your view. It literally overwhelms a telescope's field of view. Instead of seeing the beautiful coma and tail, you see only a small portion of the coma. For the full effect of a comet you are better off using a pair of binoculars or your own eyes.

Observing Tip 4: Know where and when to look.

There are many comets with known orbits. However, most of them are distant and very faint, which means you have to hunt for them. New comets are discovered all the time. The trick to observing a bright comet is to keep your eyes and ears open to the media. If a comet is going to put on a good show, the media will let you know when and where to look.

OBSERVING FOR THE FUN OF IT

PART II

OBSERVING THE SKY WITH A LITTLE HELP

7

Binoculars For Sale?

Many people already have a pair of old binoculars lying on a shelf in some closet. If you're one of them, you might want to dust the binoculars off and see how they rate, using the information in this chapter. If you are thinking about buying new binoculars, read on for some simple but important tips. Binoculars can be very useful when it comes to checking out the sky.

How Much Will This Cost Me?

One of the first things people consider when they want to buy something is the price. Binoculars are no exception. As is true of many products, binoculars come in a wide range of shapes, sizes, colors, and (most importantly) prices. Where do you begin? Well, the choice of color is completely up to you. For help with everything else, read on.

Buying binoculars is like any other purchase you may make—you get what you pay for. True, you can probably purchase a pair of binoculars for only $20 at the local swap meet. Department stores may also have bargain binocular deals beginning at $29.99 and up. However, this is not necessarily the best way to begin a beautiful astronomical relationship.

To insure that you get a good pair of binoculars, plan to spend at least $100. Now, don't faint until you read the rest of this section. Spending $100 up front for a good pair of binoculars will save you money and frustration in the long run. What you are really paying for with a good pair of binoculars is not only quality, but an extended warranty as well.

Why would you want a warranty? Even if you treat your binoculars with the best of care, you may eventually need to send them back to the factory for maintenance or repair. The warranty allows you this option. Cheaper binoculars are not only made of cheaper materials, their warranties are very limited. You may have only a few months of good observing before you start seeing double or some other problem crops up. Then you are stuck.

What Are You Buying?

After you recover from sticker shock, it's time to decide what exactly you'll be spending your money on. All binoculars are shaped pretty much the same way—they're two long tubes hooked together in the middle. What can vary in binoculars is the size—not only the length of the tubes, but the diameter of the largest lenses, as well.

When you first start looking for a pair of binoculars, you will be inundated by numbers: 7x35, 8x40, 10x50, etc. What these numbers stand for is very simple. The first number is the amount of magnification the binoculars produce (in other words, how much closer an object will appear). The second number is the diameter in millimeters of each of the two lenses at the objective end of the binoculars (the big end, not the eyepiece end). The larger the number, the bigger the lens and the more light it gathers. For astronomy, this feature is very important. After all, it's dark outside and the objects you will be looking at are very faint. The more light you can gather, the better you will be able to see.

The most common size for binoculars is 7x35. These binoculars magnify an object 7 times and have lenses that are 35mm in diameter. They are okay for everyday use. But for astronomical observing, you will want a bigger lens diameter. 7x50 or 10x50 binoculars will work quite well for first-time astronomers.

Observing Tip 1: Have a steady base.

Day or night, the key to success when it comes to observing with binoculars is the ability to keep them steady. Holding the binoculars in your hands gives you the freedom to look in any direction you wish. However, without a steady base you may get dizzy trying to observe an image that bounces around with every breath you take. By bracing your elbows on the hood of your car, a tree limb, or the shoulder of a friend, you are providing a steady base for the binoculars. Objects are much easier to observe when they are motionless in the binoculars.

Observing Tip 2: Focus on a star.

When you begin your observing run, start by observing a star. Stars are good objects to focus on because they appear as very distinct tiny dots when they are focused. You want to have the smallest star image possible for the sharpest focus. Focus first with the eye that has the adjustable eyepiece. Adjust the eyepiece until the star is in sharp focus. Then, using the main focus adjustments for the binoculars, focus again. Move the focus in and out until the image is at its sharpest. Also adjust the binoculars for the space between your eyes by bringing the binocular tubes closer together or farther apart.

Observing Tip 3: Observe where it's dark.

When you are observing the Moon and bright stars, your location doesn't really matter—as long as you can see the sky. However, once you begin your quest for fainter objects, the darker the skies, the better. Streetlights are great when you are driving in an unfamiliar city, but they are an astronomer's curse when it comes to observing.

Maintenance

Like anything else, your binoculars will last longer if you take care of them. By establishing a simple routine after every observing run, you can keep them in great shape for a long time.

Store your binoculars in their carrying case. If they didn't come with a carrying case, store them in an airtight plastic bag. The case will protect them from any bumps and bruises as well as dust and moisture. A plastic bag will at least keep out the dust and moisture.

Cover the lenses when not in use. If your binoculars come with a set of lens covers, use them. If you get a scratch on a lens there isn't a whole lot you can do about it.

Keep them clean. Most binoculars come with a cloth that you are supposed to use for cleaning. However, after months of being hauled around in the bottom of a carrying case, the cloth will be dirtier than the binoculars. Any dust or dirt trapped in the cloth may scratch the lenses when you try to clean them.

So instead of the cloth, use a camera brush or a bulb-type syringe to blow air over the lenses. Both can be purchased at a camera store. If all else fails, you can use lens-cleaning solution and cotton swabs. The cleaning solution can also be purchased at a camera store. Pour a small amount onto the lens and gently clean it with the cotton.

If you start to see double: After you have been observing for a while you may start to have trouble focusing or see a double image when you look at a star. The double image is not the result of your advancing age, but a result of the optics in the binoculars going out of alignment.

Binoculars have two separate lenses that collect light (one for each eye). Inside the binoculars is a series of prisms that bend the light so your eyes pick up only one image. When these prisms are out of alignment, you begin to see two images instead of one. The technical phrase used to describe this binocular ailment is "out of collimation."

This ailment is rather common with binoculars that have had a lot of use or abuse. Usually, binocular manufacturers guarantee lifetime collimation. If you do run into a problem, take the binoculars back to the place you bought them and ask the salesclerk for help. They may have to send the binoculars back to the manufacturer. Once you have made your purchase and are ready to enjoy the night sky, turn to Chapters 10 and 11 for a list of objects to enjoy.

8

How to Buy a Telescope

The universe is a fascinating place, filled with stellar nurseries, exploding stars, planets many times the size of Earth, and distances that defy our comprehension. A telescope can be your key to entering this fascinating final frontier, but getting your hands on one may not be easy. Trying to sort through all the information involved with buying a telescope can be more intimidating and confusing than the vast reaches of space you hope to explore.

This chapter tries to take away some of that confusion by explaining the basic things to look for when buying your eye to the sky. You don't need a degree in math or physics to make an intelligent purchase. In fact, armed with just a few basic tips, you can change a confusing and expensive task into a fun experience and get a good, usable telescope in the process.

Types of Telescopes

First of all, what is a telescope? The dictionary defines it as "a tubular optical instrument for viewing distant objects by means of the refraction of light rays through a lens or reflection of the light rays by a concave mirror."

Simple? Let's try that again. A telescope is something you use to get closeup views of distant objects. How was that? Better? Unfortunately, when you walk into a store or flip through a telescope catalog, you are going to find more things like the first definition than the second, so let's go back to the first definition and try to figure out what it means.

There are two basic types of telescopes: *refractors* and *reflectors.* These are the two types mentioned in the definition. You can use these terms to

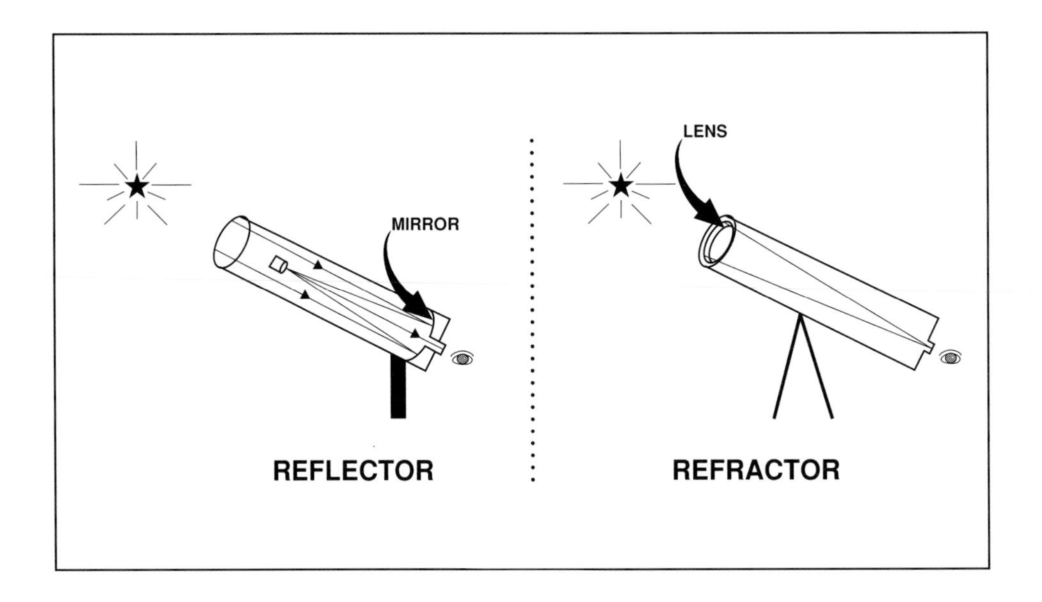

impress upon salespersons that you know a thing or two about telescopes. A refractor uses a lens to gather light and do its magic. A reflector uses a mirror. An easy way to remember: mirrors reflect—reflectors use mirrors.

Your next question may be, "Which is better?" This is a simple question, but it has a rather complicated, drawn-out answer. One type is not really better than the other. It all depends on what kind of objects you want to look at. While one telescope type may see some objects better than the other, the viewing differences between the two types are not really noticeable with small telescopes.

One difference you will notice between the two is cost. With your first telescope, cost is usually a deciding factor. Small refractors are the least expensive, and this is why most of the small telescopes you'll find in retail stores are refractors. However, the tips given in this book are relevant to either type.

Things to Keep in Mind

There are four things to consider when buying your first telescope: cost; the size of the lens or mirror (the bigger, the better); magnification (forget it); and a sturdy tripod. Of course, after you become a more experienced stargazer there will be other things to consider, but for now let's keep it simple.

Cost. Prices fluctuate and it is hard to quote exact dollar amounts, but there are some general guidelines. If you, your child, husband or wife, father or mother are really interested in astronomy, it would be wise to invest your money in a nice telescope instead of buying the cheapest one you can find. The quality of anything under $100 will frustrate the future astronomer and probably end up in the closet after one or two failed attempts at observing.

Prices increase with the size of the telescope. "Size" refers to the diameter of the largest mirror or lens used, not the length of the telescope's tube. The better telescopes found in department stores have at least a 2-inch (50mm) lens. The bigger the main lens or mirror, the better. Most people start out with a 2- to 3-inch (50 to 75mm) telescope. If you ask some simple questions and pay attention to some basic details, you can get a fairly good beginning telescope for between $200 and $300.

The bigger around, the better. When you walk into a store's camera section (where the telescopes are usually displayed) one of the first things you see is a huge sign claiming, "Magnifies 600 Times!" As tempting as that may sound, your best bet is to turn the other way.

More important than magnification is something you can identify without all the fancy signs or helpful sales clerks. All you need to do is look at the telescope and remember: The bigger around it is, the better.

The diameter of a telescope, or more precisely the diameter of the largest mirror or lens in the telescope, determines how much light it gathers. *Light-gathering power* is much more important than magnification. For this to make sense, let me break away and give an example that will seem completely unrelated to the topic.

Imagine that you are dying of thirst in a desert. All of a sudden, it starts to rain. Quickly you grab the two closest containers and put them outside. One is an empty 2-liter pop bottle, the other is a coffee mug. If it only rains for two minutes, which will collect the most water?

This is not a math problem. Just use common sense. The pop bottle is bigger, but it has a smaller opening. Not much rain will get into the bottle through the small hole. The coffee mug will do a better job of collecting water and quenching your thirst because it has a bigger opening. The bigger the opening, the more water collected.

What does this have to do with telescopes, you ask? Well, telescopes don't collect water, but they do collect light. The bigger the mirror or lens in the telescope, the more light it will gather. With more light, whatever you look at will be brighter. This is important because most of the objects you will observe are extremely faint. The more light you collect, the more details you will see.

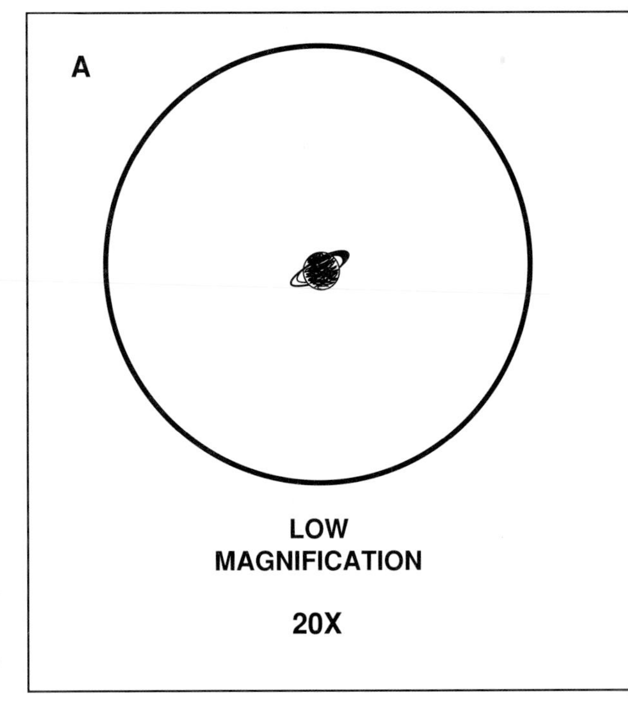

Magnification is not important. Magnification, or power, is not the important item to consider when buying a telescope. (In advertising, the words *magnification* and *power* are used to describe the same thing.) In fact, magnification should be one of the last things you think about.

When you magnify something, you don't change how much light is coming into the telescope. (To do that, you would have to change the size of your main lens or mirror.) Instead, you change how big that object appears. You do this by spreading out the light that is already there.

Drawing A shows Saturn through a telescope using low magnification (around 20 times.) The image is small, but bright. Several details can be seen.

Drawing B depicts Saturn through medium magnification (around 80 times.) It is larger, but you start to lose some of the detail. Remember, you start out with the same amount of light. The more you magnify, the more that light gets spread out.

Drawing C shows Saturn through high magnification (about 150 to 200 times.) The image is larger, but you have lost a great deal of detail and the image itself is much fainter.

So high magnification is not always a plus. Highly magnified images are large but very faint and difficult to see. For a small telescope, any object that is magnified over 250 times is almost impossible to see. Your best bet is to stay below 100 power when you are getting started.

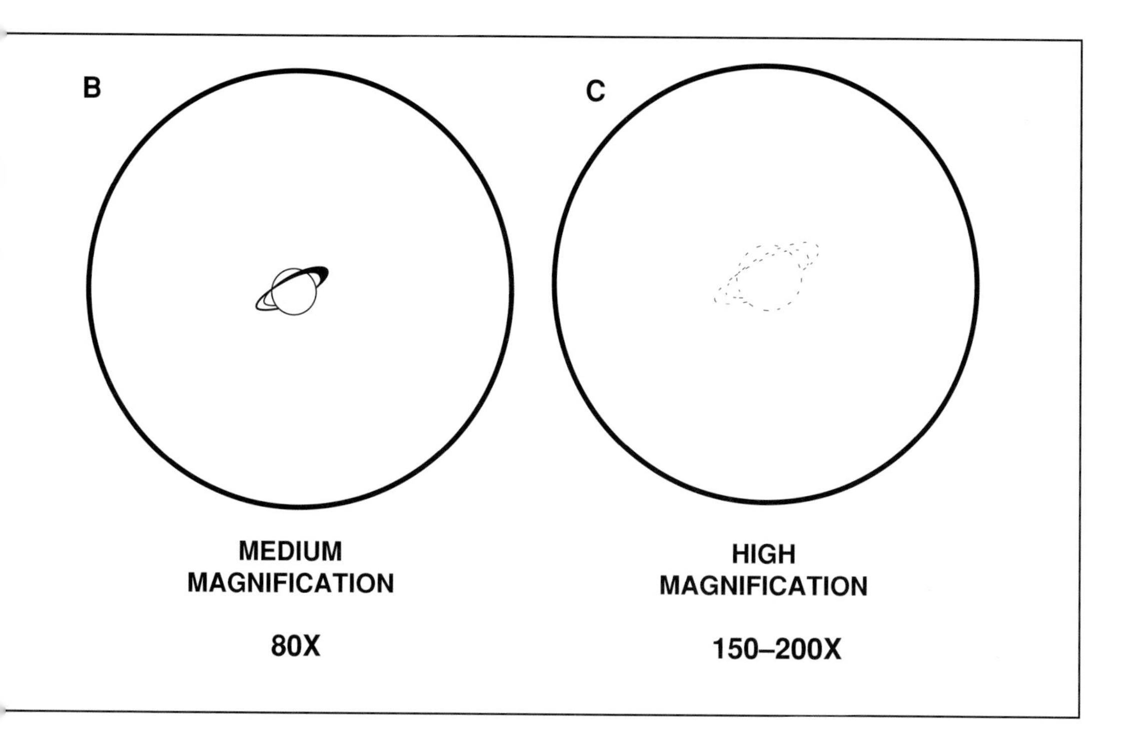

A steady tripod. Another item to consider besides money, light-gathering power, and magnification is a steady tripod. Choosing a steady tripod is just as important as ignoring the magnifying power, but you never see this fact plastered on sale signs. In fact, most people don't even consider the tripod when buying their telescope. They don't realize the importance of a tripod until it's too late.

For example, imagine you have spent 15 minutes searching the skies for Saturn. Finally, Saturn and its beautiful ring system swing into view. As you gently focus in on our distant neighbor, your dog decides to see what all the excitement is about. With one wag of his tail he bumps a wobbly tripod leg and sends your telescope pointing toward never-never land. You now face another frustrating search and a confused dog.

To avoid the above scenario, a tripod should be very sturdy, with adequate supports. If you are shopping for telescopes in a store, you can actually test the sturdiness of a tripod. Aim the telescope at an object across the room or outside (do not aim a telescope at the Sun). Then gently bump the tripod. See how much force it takes to knock the telescope off your object. Also try gently tapping the top end of the telescope itself. It should be able to withstand a gentle nudge. You don't want a tripod that will move in a gentle breeze.

While you are checking how sturdy the tripod is, also check how firmly the telescope is mounted to it. After you have nudged the tripod a few times, take hold of the telescope itself and try moving it while you are looking at a distant object. If the telescope is mounted firmly to the tripod there should be no movement other than those the telescope adjustments allow.

A few more items to consider. While you have the salesperson's attention, ask how portable the telescope is. If you plan to travel to dark skies, make sure the telescope will fit in your car. Also, make sure it isn't too heavy to carry.

Take some time to experiment with it. How easy is it to use? Are the adjustment knobs easy to turn? When you tighten them, can you still move the telescope? (You shouldn't be able to.) Are they easy to reach?

These types of questions can be answered in the store before you buy. Play around with the movement of the telescope and see how it feels. It is important not to be intimidated by something you are planning to buy.

A brief summary. Know how much money you want to spend. The bigger the diameter of the telescope, the better. Forget magnification or power. Check for a sturdy tripod. Ask how portable the telescope is. (Can you lift it?!) Armed with these few tips, you should feel prepared to purchase your new telescope.

Accessories

Now that you have a pretty good idea of what to look for in a telescope, get ready for some more technical jargon. Usually, when you buy a telescope, you will be offered all kinds of accessories. Many kinds of eyepieces, a Barlow lens to increase magnification, a Moon filter, a solar filter, and an elbow or a right-angle piece are just a few of the items that might be thrown at you.

Keep in mind that these accessories are exactly what they sound like, accessories. It's not necessary to have them. The telescope will work fine on its own, but feel free to consider them.

Accessories will enhance your telescope's performance. Some people purchase their accessories with the telescope, others wait to see what they'll need. What type of accessories to buy and when to buy them are really up to the individual. Now let's sort through some of these accessories and see what you may or may not need.

Barlow lens. Some telescopes come with a Barlow lens. It is placed between the telescope and the eyepiece. A Barlow lens doubles the amount of magnification of your eyepiece. With small telescopes, Barlow lenses are not really recommended. Remember, when you magnify an object, you don't increase the amount of light coming in, you only spread out what little light there is. A Barlow lens spreads the light out so much that the image is very faint and difficult to see. It's also difficult to look through and requires some practice before you get the hang of it.

It's a good idea to put the Barlow lens aside until you've observed with your telescope and are comfortable with the way it works. Then try using the Barlow lens on bright objects such as the Moon or Jupiter.

Elbow, or right-angle viewer. This piece is shaped like an L. The eyepiece fits into one end of the L. The other end is placed in the telescope. Inside the elbow, or right angle, there is a prism or mirror that takes the light from an object and reflects it 90 degrees. This will be very useful when you are trying to look at something straight overhead. A right angle isn't necessary, but it really makes viewing more comfortable.

Eyepieces. The eyepiece of a telescope is very important. After all, an eyepiece is half of your telescope's optical system. On some small telescopes the eyepiece is permanently attached. There isn't much you can change with that setup. If that is the case with the telescope you are considering, skip to the next accessory. If you can change eyepieces, read on.

On some telescopes you can remove the eyepiece and switch to another. When you change the eyepiece, you are changing the magnification of the telescope, giving you a little more flexibility.

An eyepiece can be identified by a number stamped on its side. This number can be anywhere from 4 to 40 and will always be in millimeters (mm). This number is the focal length of the eyepiece. Most people refer to an eyepiece by this number. Eyepieces with different focal lengths magnify different amounts. The tricky thing to remember is that the smaller the number, the more it magnifies. For example, a 6mm eyepiece may magnify 150 times, while a 20mm eyepiece magnifies 45 times. Just remember that it's backwards—the bigger the number on the eyepiece (the focal length), the less it magnifies.

Eyepieces can be any size from 4mm to 40mm and larger (4mm magnifying the most, 40mm magnifying the least). The most useful eyepiece sizes for a small telescope are between 6mm and 40mm. The amount a specific eyepiece will magnify varies from telescope to telescope.

When you are starting out, stick to the larger numbered eyepieces with less magnification. Objects will be easier to find, so you won't be discouraged.

Another note: Keep in mind that the more you magnify an object, the more you magnify the atmosphere between you and the object. So if the sky is hazy, you magnify the haze as well as whatever you are looking at. If it is cloudy, you won't be able to see anything.

When you try the high-power eyepieces (12mm to 4mm), your best bet is to stick to observing bright objects such as the Moon and planets. Because the high-power eyepieces magnify so much and look at such a small area of the sky, you may find it difficult to see anything at first. To locate the object you want to observe, use a low-power eyepiece first. Things will be easier to find with less magnification. When you have the object centered in the telescope, very carefully change to the higher eyepiece. Try not to move the telescope while changing eyepieces or you will be back where you started—lost in space.

Finderscope. On most telescopes a finderscope is located on the side of the main telescope. The telescopes that don't have finderscopes are usually the smallest telescopes available. Any telescope over 2 inches in diameter really should have one.

Finderscopes do just what their name suggests—they help you find things. Finderscopes are mini-telescopes, and the amount they magnify varies with the telescope, but they are not nearly as powerful as the telescope they are attached to.

Through the finderscope you can see a larger area of the sky, which will help you locate the object you want to observe. Keep in mind: The larger the telescope, the greater the need for a finderscope.

Moon filter. Some telescopes come with a small Moon filter that may screw into the eyepiece. If your telescope doesn't have one, don't worry. A piece of cardboard covering part of the end of the telescope works just as well. If you can't find any cardboard, put on your sunglasses and take a peek. Trust me. It may sound crazy, but sunglasses do work as a Moon filter.

The only thing a Moon filter does is cut down on the amount of light that passes through the telescope. Many times the Moon is too bright to see much detail. With a Moon filter or even just a piece of cardboard blocking out some of the light, you can see more detail on its surface.

Something to remember: If you use the Moon filter to observe the Moon, be sure to take it off when you try to look at something else. Otherwise, the filter will cut off the light from the fainter stars and you may not be able to see anything.

Slow motion controls. Every telescope has two sets of controls—one to move the telescope back and forth, the other to move it up and down. These controls are necessary for large movements, such as moving the telescope from one object to another. For making small adjustments after you have found an object, most telescopes have *slow motion controls.* While you are checking over your potential purchase, ask the salesperson if the telescope has slow motion controls. Most, but not all telescopes have them. They are handy to have, since they make small adjustments easier. Slow motion controls usually add about $50 to $75 to the price of a telescope. If you have the money, I would recommend investing in them. However, if you are just starting out, they're not absolutely necessary.

Sun filter. Some telescopes come with a small Sun filter that screws into the eyepiece like a Moon filter. If you have one of these filters, throw it away. These filters are made of smoked glass and can shatter if they get too hot. If a filter shattered while you were looking at the Sun, the magnified sunlight would burn your eye before you could pull away. You would be blind in that eye.

The Sun is a dangerous object to look at, and once your eye is burned there is nothing you can do about it. So play it safe—don't look at the Sun unless you are with someone who knows what he or she is doing and has special filters that are placed over the end of the telescope. Even though the

Sun is a fascinating object, it is very dangerous to look at and is not worth losing your eyesight over. If you're not sure, don't!

A Brief Accessory Summary

Accessory	Recommendation
Barlow Lens	Not recommended
Elbow/Right Angle	Recommended for more comfortable viewing
Eyepieces	Larger focal lengths for beginners
Finderscope	For 2-inch and larger telescopes
Moon Filter	Can use substitutes
Slow Motion Controls	Nice, but not mandatory on first telescope
Sun Filter	Absolutely not

Now What Do I Do?

You've done your research, you've shopped around, you have quizzed the salesclerk with intelligent questions, and you've purchased a brand new telescope. Now what? Place it in the corner and stare at it? No. Now is the time to have fun. In the beginning, it's best to have fun during the day when you can see what you are doing.

When you take your new toy out of its package, watch for small parts. You may have to attach some things yourself. Follow the instructions that come with the telescope, making sure everything is tightened securely.

When you have everything together you may want to rush outside and immediately begin exploring the heavens. If you do, you may be disappointed. Like many other things in life, observing through a telescope is not as easy as it looks. Take time to practice on some easy targets before you aim for the great unknown. Just a few minutes of practicing will make your first night out much easier.

First try the telescope during the day when you can see what you're doing. How does it move? How do you adjust the tripod? How do you change eyepieces? Try closing your eyes and finding the focus knob or the adjustment knobs. Remember, you are going to use the telescope at night, so you won't be able to turn on a light every time you have to make an adjustment.

After you are familiar with the various adjustments, try finding a few objects in your neighborhood. As you gaze through your telescope for the first time you will notice that the image you see is upside-down. Don't worry—

your telescope isn't broken. With most astronomical telescopes the image is upside-down. It has to do with the way the lenses magnify. This may make things confusing as you observe objects during the day, but at night you won't notice it.

While you practice finding trees, mailboxes, streetlights, and other objects, this is a good time to adjust the finderscope if you have one. Usually it isn't aligned with the main telescope when you bring it home.

Aligning a Finderscope

Center the main telescope on something obvious like a corner of a neighbor's roof, the top of a flagpole, or a streetlight. Make sure it is easy to identify. A tree limb is okay as long as you can tell exactly which branch you are centered on. Then look through the finder. If the center of the finderscope is on the same spot as the center of the main telescope, you are all set. If not, it isn't difficult to fix.

Slightly loosen one of the screws holding the finderscope. Look through the main telescope and make sure you are still centered on your object. Then without moving the main telescope, look through the finderscope. Carefully tighten the screw opposite the one you loosened. As you tighten the screw, the finder will move. Be sure to look through the main telescope several times as you are adjusting the finderscope. You may have to adjust the screws several times before the two telescopes are aligned on the same object.

It is much easier to align the two telescopes during the day, but if you need to do this at night, aim for a distant street lamp or a radio tower. After the two are aligned, make sure all the screws are tightened down. Now you're ready for the first night out.

Where and What Do I Look At?

The Sun is going down, the skies are clear, the wind is calm. All good signs for a wonderful evening of exploration. Your telescope is all prepared for the evening to come, so before you go outside, make sure you're prepared, too. Dress warmly. Even on muggy summer evenings you may be surprised at how cool a slight breeze can be. Brisk autumn evenings can become frigid after you stand in one spot for an hour. During fall and winter nights you may want to wear a couple of pairs of socks. Observing requires moving the telescope around quite a bit, but the observer doesn't get much exercise. Doing jumping jacks during observing breaks does help a little to restore circulation. Also, don't forget the insect repellent.

When you are as prepared as your new telescope, you are ready to tackle the heavens. Take your telescope outside to a dark area away from the bright lights of the city. Make sure the sky is clear. No matter how big your telescope is, it can't see through clouds. You set up your telescope with practiced ease, put in an eyepiece, aim at the sky and you see . . . nothing. Instead of thousands of stars or a brilliant cratered Moon, chances are the first view through your telescope will show nothing. Don't be discouraged. Observing through a telescope, like anything else, takes practice.

The best object to start with is one of my personal favorites, the Moon. Even though it is really easy to see in the sky, you may find the Moon difficult to locate in your telescope. Practice aligning your telescope on the Moon before you move on to smaller, fainter objects.

If the images you see appear blurry as you observe them, check your focus. Turn the focus knob back and forth until the image is crystal clear. After you have mastered the Moon, you can explore the other fascinating objects in our universe. Planets, star clusters, nebulae, and galaxies all await your viewing pleasure.

One more thing—don't expect to see a lot of color or extremely large, brilliant galaxies and nebulae. Pictures you see in magazines and books were taken with very large telescopes or, in the case of the planets, spacecraft flying close by. A long-exposure photograph can show much more detail than the human eye can see.

The objects in your telescope may at first seem very faint and fuzzy with very little detail and no color. Even a star will only appear as only a single pinpoint of light, no matter how much magnification you use. After you have observed for a while you will begin to make out subtle differences in those fuzzy gray clouds that the books call galaxies, star clusters, or nebulae. Pretty soon, finding that elusive, extremely faint nebula will become your challenge.

Binoculars vs. Telescopes

After reading all about binoculars and telescopes, you may be wondering which is best for you. When most people think about astronomy, they think telescopes. However, for some people entering the world of observational astronomy, binoculars may be the better choice. To make your decision a little easier, I've compared the two choices and have come up with a list of advantages that one has over the other.

Advantages of Binoculars

Easier to see through. Binoculars are easier to see through because you are using both eyes to observe. Children may have better luck seeing objects through a pair of binoculars.

Easier to find objects. Binoculars have a wider field of view, which means you are looking at a larger area of space. This wider field of view makes it easier to find those tiny objects. You can usually get by with aiming the binoculars in the general direction of an object and still be able to find it. With the narrow field of view of a telescope you must be very precise with your pointing. If not, the object may be outside your field of view and you won't see it.

For example, imagine that you are looking for a particular crater on the Moon. Through a pair of binoculars you will be able to see the entire Moon and can pick out the crater with ease. Through a larger telescope you may be able to see only a few craters at a time. To find the one you want you will have to scan back and forth across the Moon's surface.

Advantage of Telescopes

Magnification. Telescopes gather more light than binoculars. The more light you gather, the more magnification you can use. Magnification is really important if you are interested in viewing the planets. A telescope is required to see the cloud bands on Jupiter or resolve the rings of Saturn.

Deciding which to buy depends entirely on the individual. Before you buy anything, look over both options, then purchase the one you'll feel comfortable using. If you don't feel comfortable with the workings of a telescope you may want to start out with binoculars. If you really want to see the planets and the telescope appears to be user-friendly, start with the telescope.

These tips are meant for those of you just getting started. If you catch the observing bug and get hooked you can always upgrade your equipment to a larger telescope. Remember, observing is supposed to be fun. Take it slow at first. Whether you have binoculars or a telescope, get used to how they work before you try observing at night. Being prepared makes for a much more enjoyable observing run.

10

Observing The Moon And Planets

What Will You See on the Moon?

Craters, craters, a few smooth, dark areas, and more craters. No, you won't be able to see any flags or footprints left behind by the astronauts. No telescope on Earth, or in Earth orbit for that matter, can see that well.

Craters. Why craters? Because any and every piece of space junk aimed at the Moon hits it and leaves a crater on its surface. Most of the stuff aimed at Earth burns up as it tries to pass through our atmosphere. And even if a crater is formed on Earth, the wind and rain wash it away before long. The Moon doesn't have an atmosphere to protect it, so it gets hit by everything aimed in its direction and it has a lot of craters to show for it. Not only does the Moon end up with a bunch of craters, but once a crater is formed it is stuck on the Moon for millions of years. In fact, the only way a crater disappears on the Moon is if another crater forms on top of it. With no atmosphere, there is no wind or rain to wash it away.

Most of the heavily cratered regions are found near the south pole of the Moon in an area called the *heavily cratered highlands.* This region represents some of the oldest terrain on the Moon. Samples brought back by astronauts date the age of the highlands at almost 4.5 billion years. The highlands are littered with thousands of craters. With so many craters, you will find craters within craters overlapping craters.

Maria. But what about those dark, smooth areas? They only have a few craters. The dark, smooth areas are called maria (pronounced MAR-ee-uh—first syllable rhymes with car) and if you observe these dark areas carefully, you may notice that many of them have a familiar rounded appearance. Recognize that shape? You guessed it—they're craters. Maria are very large craters that have been filled in by lava from cracks in the Moon's crust (not from volcanoes—the Moon doesn't have volcanoes).

Long ago when these large craters were formed, the impacts were so powerful they cracked the crust of the Moon. Like water seeping into a basement, molten lava from the interior of the Moon seeped through the cracks. The lava didn't fill in the craters overnight. Instead, it took its time. The crater floor was covered and the lava hardened, leaving a smooth surface. The rim of each of these old, buried craters can still be seen as a ring of mountains around the maria. The Apollo astronauts brought back rock samples that date the surface within the maria at 3 to 3.5 billion years old.

Just about everything you see on the Moon is a result of something slamming into our poor little neighbor. Don't feel too sorry for it, though. Just remember, if that something hadn't hit the Moon, it might have hit Earth.

Best Time to Observe the Moon

The Moon is best observed from its waxing crescent phase through its First Quarter phase (from the time it appears in the shape of a banana until you see half of it) and again from its Third Quarter phase through its waning crescent phase. There is no best day (or night) to observe the Moon. Instead, there is a period of a couple of weeks when the Moon looks pretty good through a pair of binoculars or a telescope.

Any time the Moon is less than half full you can see more details on its surface, especially along the terminator. The terminator marks the dividing line between day and night. The shadows cast by the Sun striking craters and mountains on the Moon add a sense of depth to the lunar surface. This is not to say that you can't look at the Moon at other times, but by the time it has passed its First Quarter stage, it's almost too bright to look at comfortably.

The Worst Time to Observe the Moon

Full Moon is the worst time to see any details on its surface. For one thing, it is too bright. Look directly at a Full Moon through a telescope, and it's as if someone set off a flash bulb in your eyes. You will see blue spots for the next ten minutes. It's not dangerous to look at the Moon at this time, just uncomfortable.

Another reason the Full Moon is lousy to look at through a telescope is that there are no shadows to be seen. The surface appears washed out and flat.

The Planets

With the Moon so close and bright in our skies, you may find the planets a little disappointing. The disappointment will be even greater if you are expecting something like the pictures you see in magazines. Don't ignore the planets, though, even if they aren't as flashy as the Moon. Instead, think of observing the planets as a grand challenge.

The quest begins as you search the skies for them. After you have spotted one with the aid of a monthly planet finder, observe it for a while. The longer you observe, the more your eye adapts to what you are seeing and the more details you will be able to make out.

Gazing with Binoculars

Planet-gazing is pretty dull with most pairs of ordinary binoculars. All you see are pinpoints of light or very tiny disks. Smaller binoculars don't have enough magnification to observe any noticeable features on any of the planets—with the possible exception of Jupiter. Jupiter may appear as a very tiny disk. The disk won't be impressive, but if you can brace the binoculars so they don't move, you may be able to see some of the four largest moons of Jupiter.

These four moons are actually bright enough to see without binoculars. The problem is that they are so close to Jupiter the giant planet drowns them out. It's like trying to see a lightning bug perched on the rim of a giant searchlight. The key is to hold the binoculars extremely still. Through a steady pair of binoculars the moons appear as tiny starlike points lined up very close to Jupiter. Sometimes you see all four moons, and other times only one may be visible. Observe Jupiter for two or three nights and note how the moons change positions.

Gazing with a Telescope

Through a small telescope, you will notice features on the planets that will definitely prove there is something unusual about these wandering stars. If you have a telescope with changeable eyepieces, start with the lower-magnification eyepieces (the ones with the bigger numbers) and work your way up.

Mercury. Because Mercury is the second smallest planet in our solar system, it appears only as a tiny dot in a telescope. From our vantage point on Earth we can observe Mercury changing shape as it orbits the Sun, just as our Moon goes through phases. Since Mercury is never very far from the Sun, the only time you can see the planet is either just after sunset or just before sunrise. Either way, the sky isn't very dark and no details can be seen on its surface.

Venus. One of the reasons Venus is so bright in our skies is that a thick layer of clouds covers the planet. The clouds reflect the sunlight almost as well as a mirror. Unfortunately, they also block our view of the planet's surface, so all we get to see are the different phases Venus goes through as it orbits the Sun.

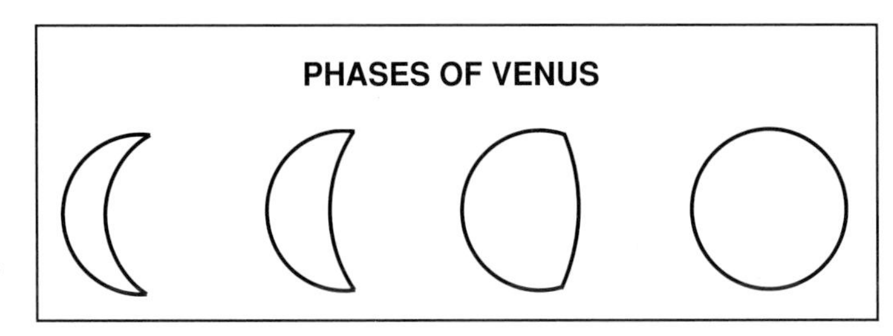

Mars. Mars is half the size of Earth and on average 136 million miles (2.28 million km) from Earth. In other words, it's rather small. Through a small telescope you should be able to make out a very tiny reddish-orange disk. The color may change as you view the planet. Sometimes you may not be able to make out any color at all. The planet is not changing—Earth's atmosphere is. The movement of air disturbs light as it passes through Earth's atmosphere, producing all kinds of colors. You may notice this effect on the other planets as well.

Jupiter. The king of the planets will put on a good show for your small telescope. Besides the four largest moons dancing around it, you should be able to see cloud bands stretching across the planet's disk. The cloud bands will appear as straight, darker lines running parallel to each other. On calm, dark nights you should be able to make out two or three different bands. With a small telescope it is doubtful that

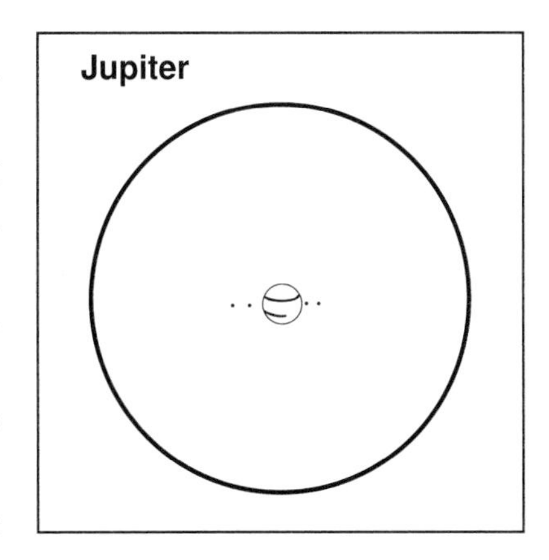

you will be able to see the Great Red Spot (a giant cyclone on Jupiter that has been around for at least 350 years).

Saturn. Through a small telescope Saturn appears to have "ears." These "ears" were first discovered by Galileo, but he didn't understand what they were. He thought he was just seeing things. It was Saturn's marvelous ring system he was seeing, and you can see it, too.

Saturn's rings are made of millions of ice particles ranging in size from a grain of sand to the size of a house. The ice reflects sunlight very well, making the rings easily visible to us on Earth. As for the planet itself, Saturn is not as colorfully marked as Jupiter. You may be able to make out one band on the disk of the planet.

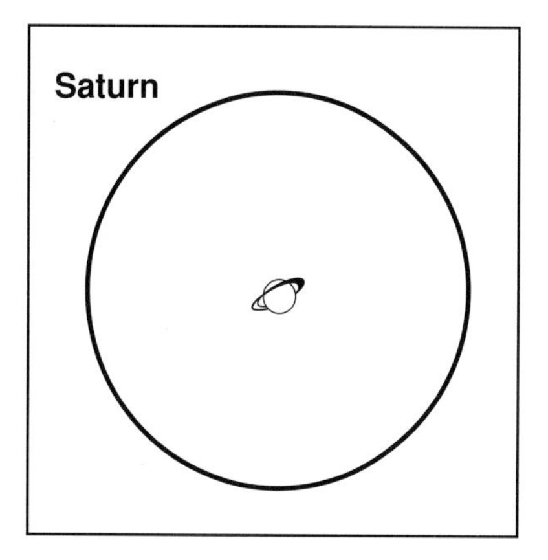

Uranus and Neptune. These two planets cannot be seen with the naked eye, which makes spotting them a little trickier than finding the other planets. Using the standard monthly planet finders, your best bet is to memorize the stars that appear to be around the planet. Then look for those star patterns through the telescope. If you can find the star pattern you know you're in the right area. Both Uranus and Neptune have just a hint of blue-green color to their tiny disks.

Pluto. Forget it. Even with a very large telescope, Pluto is nothing more than a very tiny, very faint pinpoint of light.

11

Other Objects To Look At

Now that you have conquered the Moon and planets, what are you going to look at next? With so many other objects to explore in the night sky, there are many choices. On the following pages, you will find some of the brighter deep sky objects (star clusters, nebulae, and galaxies) that can be seen through a small telescope or a pair of binoculars.

With a list of what to look at, all you need to know now is where to look. You can use the six bimonthly star maps provided or locate the objects on other star charts you may have. On each of the charts provided you'll find numbers among the stars. Each number marks the location of a possible telescope target. The numbers correspond to the objects in the section "What to Look For," below.

Use the hints given in Chapter 1, "The Constellations," to help you find your way around the star chart. The charts are very basic. One thing you may notice on the charts is that left is east and right is west. No, this is not a typographical error. The charts were designed that way. After all, where do you look when you are searching for stars (the stellar type, that is)? Straight up. If you hold the map straight over your head, with map north lined up with real north, you will see the directions are correct.

When you are ready to observe, go outside and hold the map straight overhead. Turn the map so the direction you're facing is at the bottom of the map. The stars on the bottom half of the map are what you see in the sky. Find the constellation that has your object in it. From the chart, identify the stars next to the object, then aim your telescope in that direction and observe. If at first you don't succeed, slowly sweep your telescope or binoculars back and forth across the sky in the direction indicated on the map. It may take a little while, but those objects are up there. Below are some terms you will need to be familiar with as you begin your quest for deep sky objects.

Deep Sky Glossary

Dec.: Abbreviation for *declination.* This coordinate indicates how far north or south an object is in the sky. 0° = the celestial equator, 90° = the north pole, -90° = the south pole.

Diffuse Nebula: A cloud of hydrogen gas and dust where stars are formed.

Double Star: Two stars that orbit each other.

Galaxy: An extremely large group of objects containing billions of stars, star clusters, nebulae, black holes, and much more.

Globular Cluster: A group of 50,000 to 100,000 stars bound tightly together by gravity. These clusters are found above and below the disk of spiral galaxies and are made up of very old stars.

Light-year: A unit of measurement equal to the distance that light travels in one year. At a speed of 186,000 miles a second (300,000 kilometers a second), light travels almost 5.9 trillion miles (9.44 trillion kilometers) in one year.

M: Abbreviation for Messier number. Objects found in the catalog of faint objects put together by Charles Messier, a French astronomer in the late 1700s.

NGC: Abbreviation for New General Catalog number. An extremely large catalog of faint objects.

Open Cluster: A group of 50 to 1,000 young stars formed from the same diffuse nebula. An open cluster may drift apart over time.

R.A. Abbreviation for *right ascension.* This coordinate is used to indicate how far east or west an object is located in the sky.

Spiral Galaxy: A galaxy shaped like a huge pinwheel. When seen face on, a spiral galaxy appears as a pinwheel with a bright center. When viewed edge on, you can see a thin disk with a bright ball in the center.

Star: A huge ball of hydrogen gas with nuclear reactions occurring in its core.

Mid-January (Approximately 9:00 p.m.)
Mid-February (Approximately 7:00 p.m.)

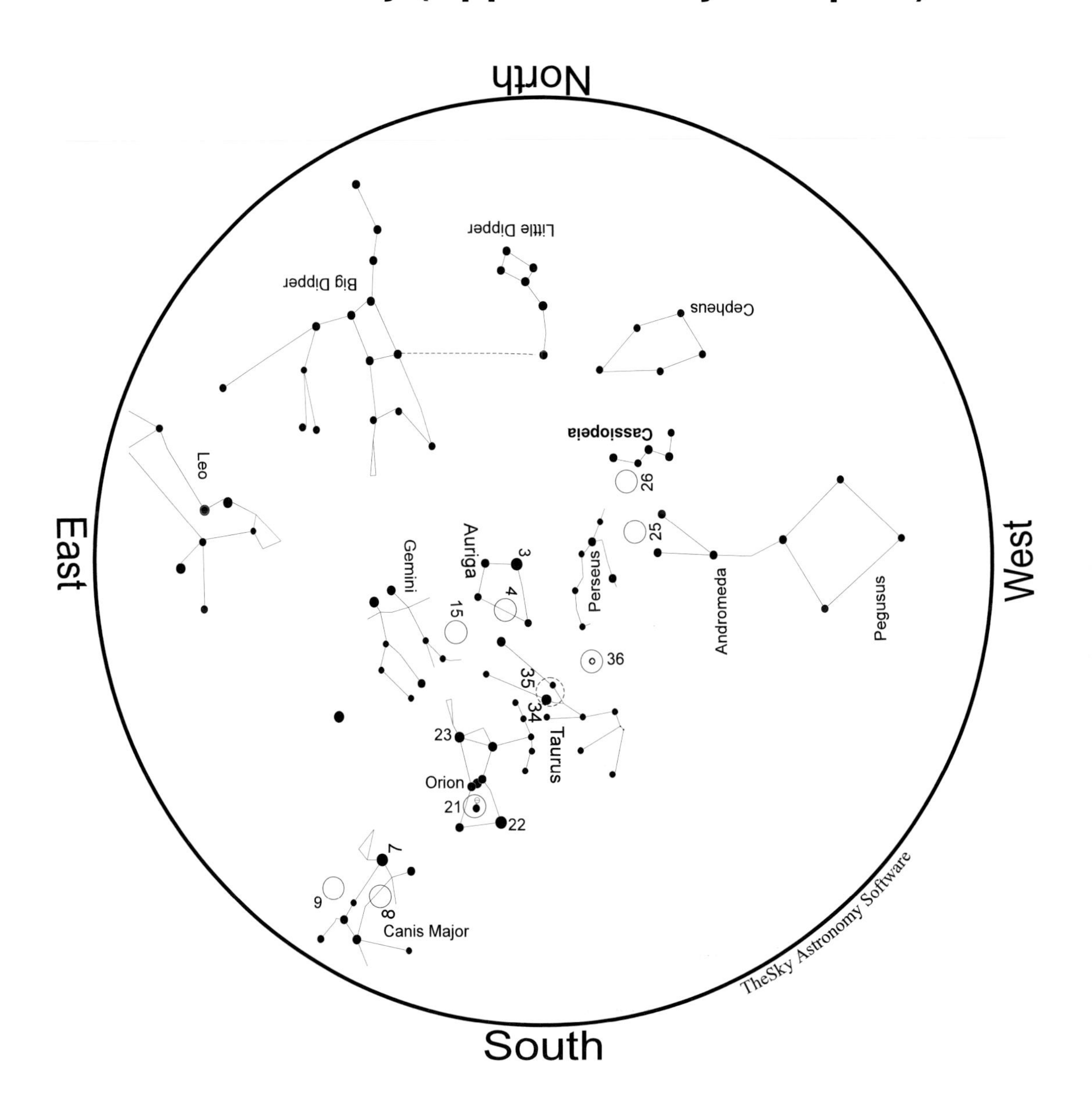

Mid-March (Approximately 9:00 p.m.)
Mid-April (Approximately 8:00 p.m.)

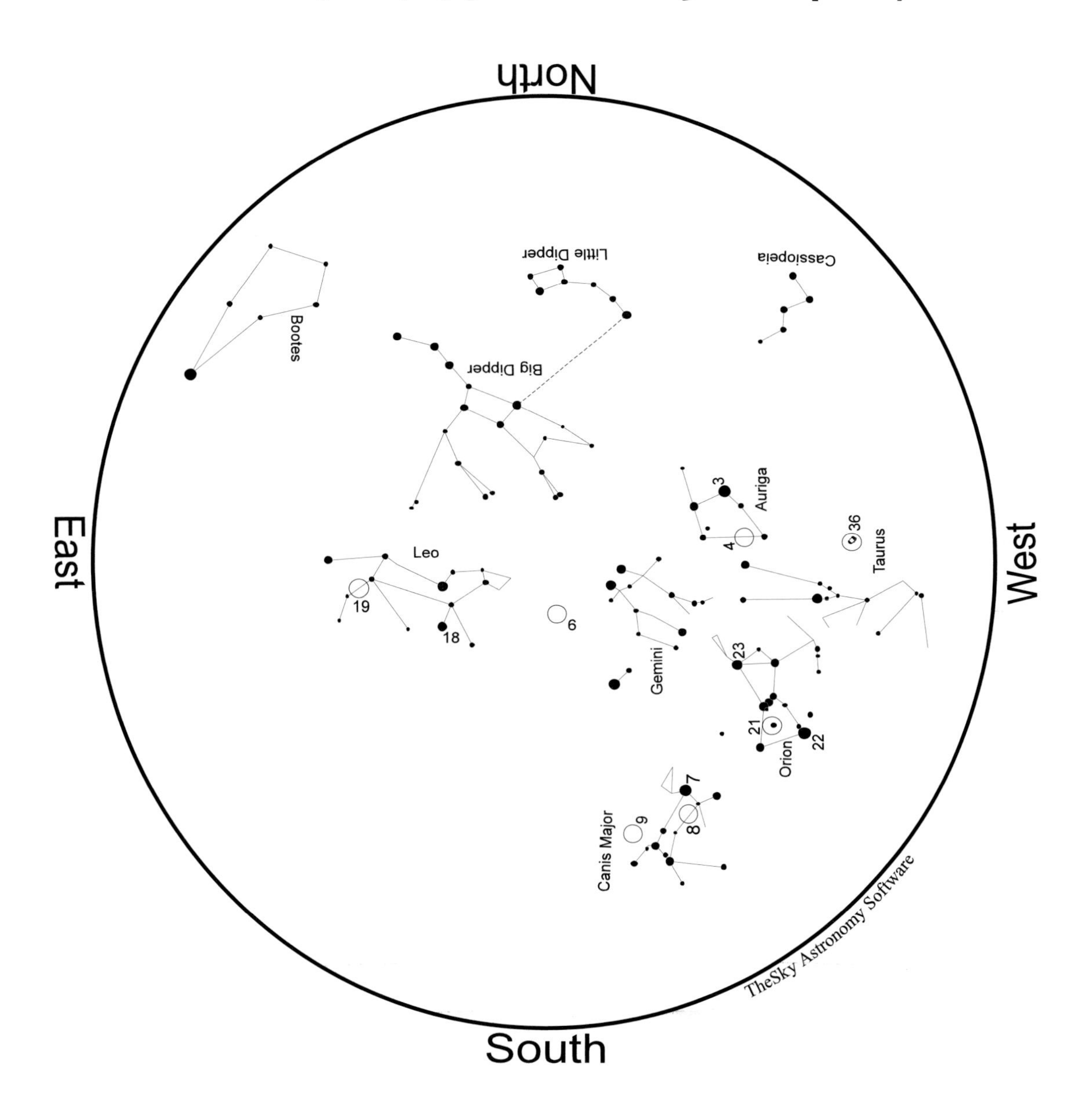

Mid-May (Approximately 1:00 a.m.)
Mid-June (Approximately 11:00 p.m.)

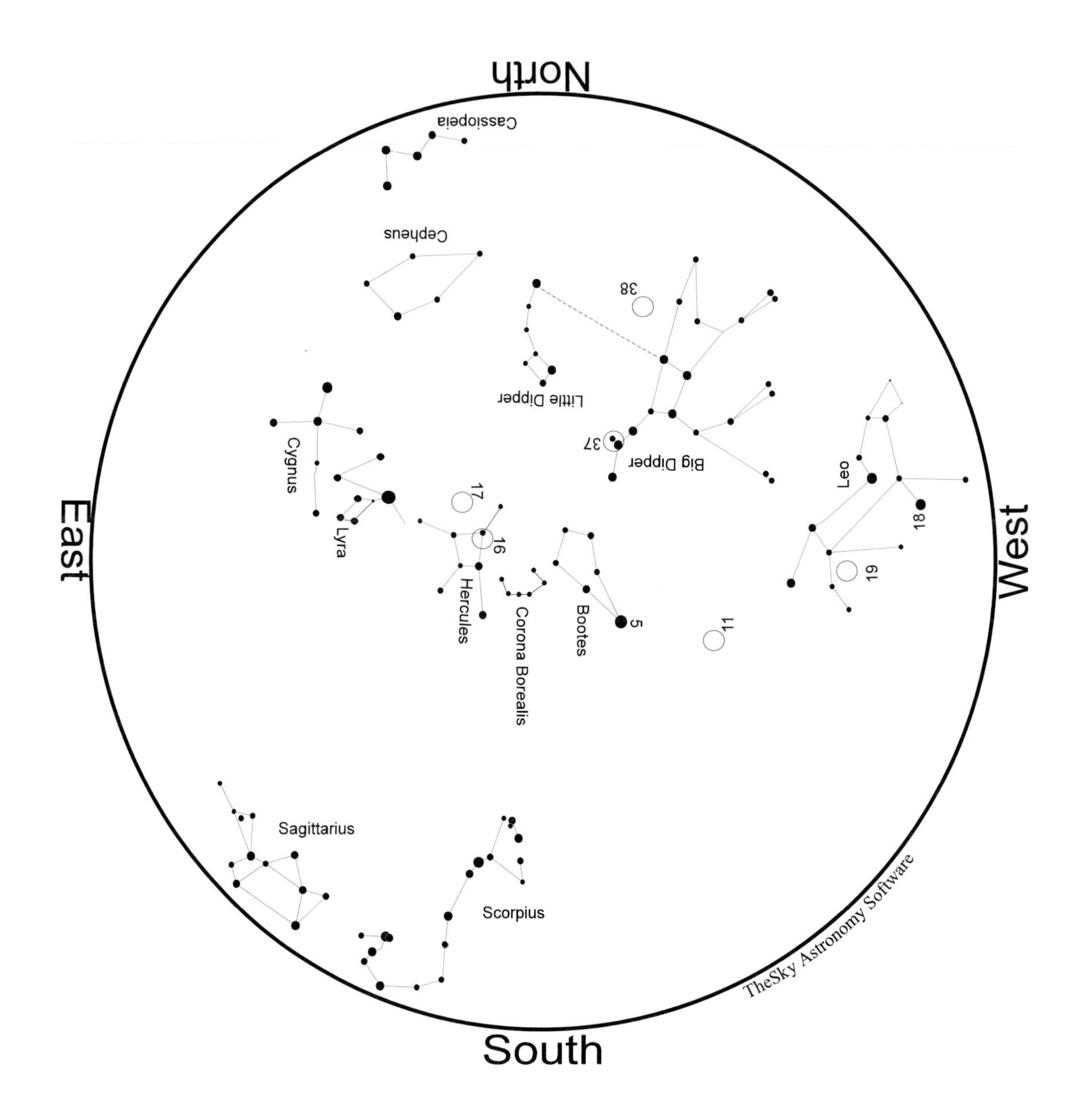

Mid-July (Approximately 11:00 p.m.) Mid-August (Approximately 9:00 p.m.)

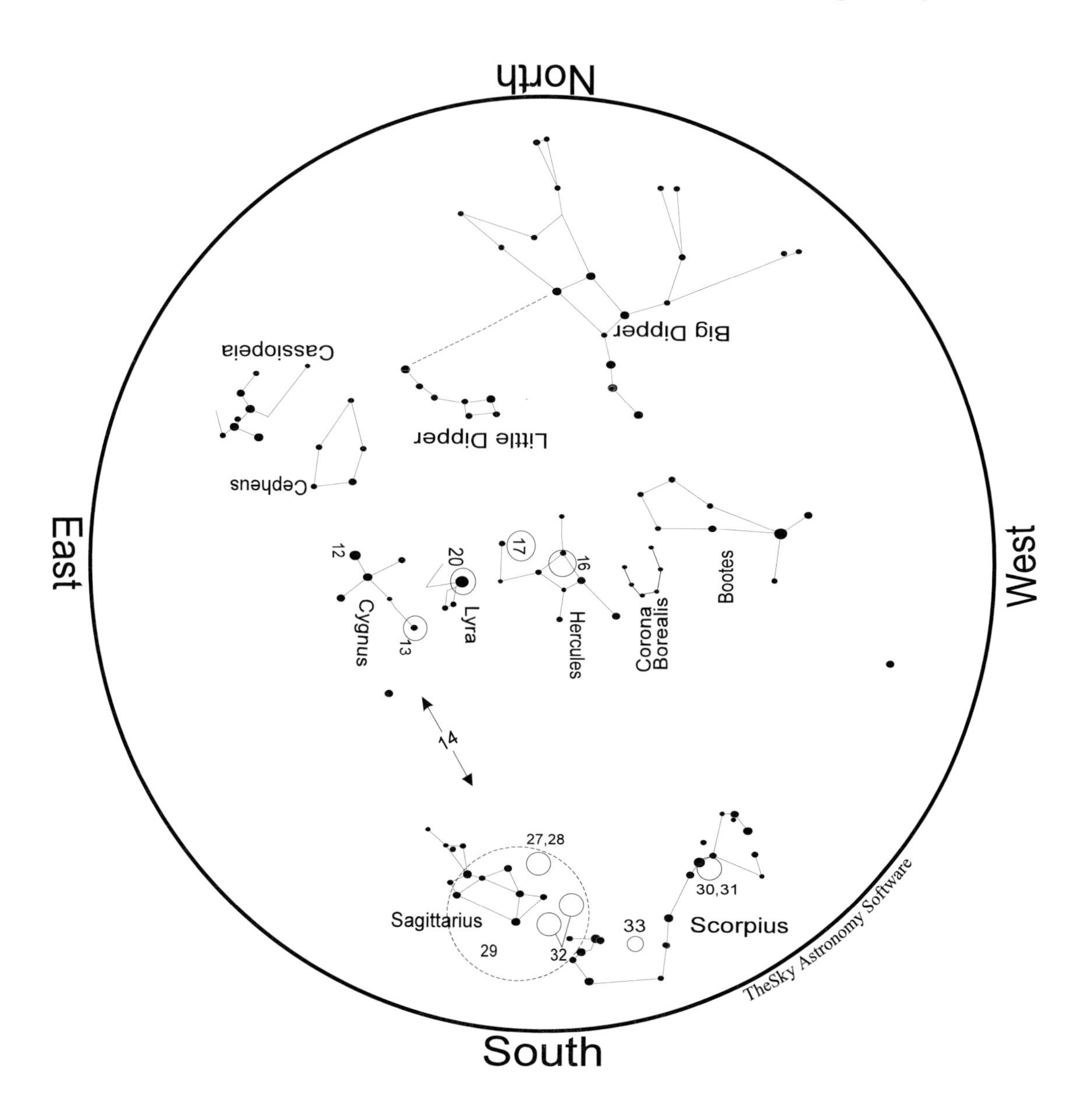

Mid-September (Approximately 10:00 p.m.)
Mid-October (Approximately 8:00 p.m.)

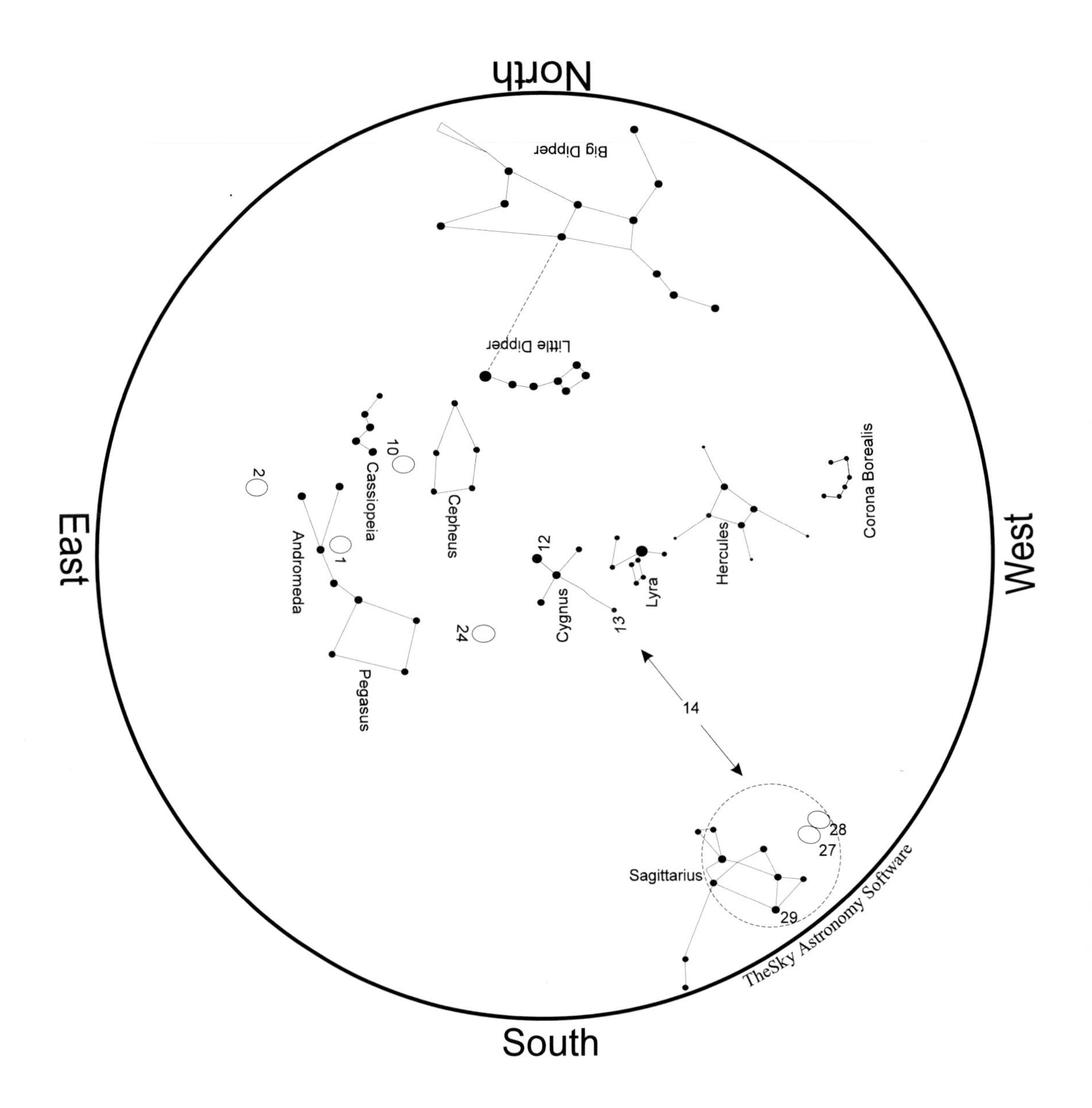

Mid-November (Approximately 10:00 p.m.)
Mid-December (Approximately 8:00 p.m.)

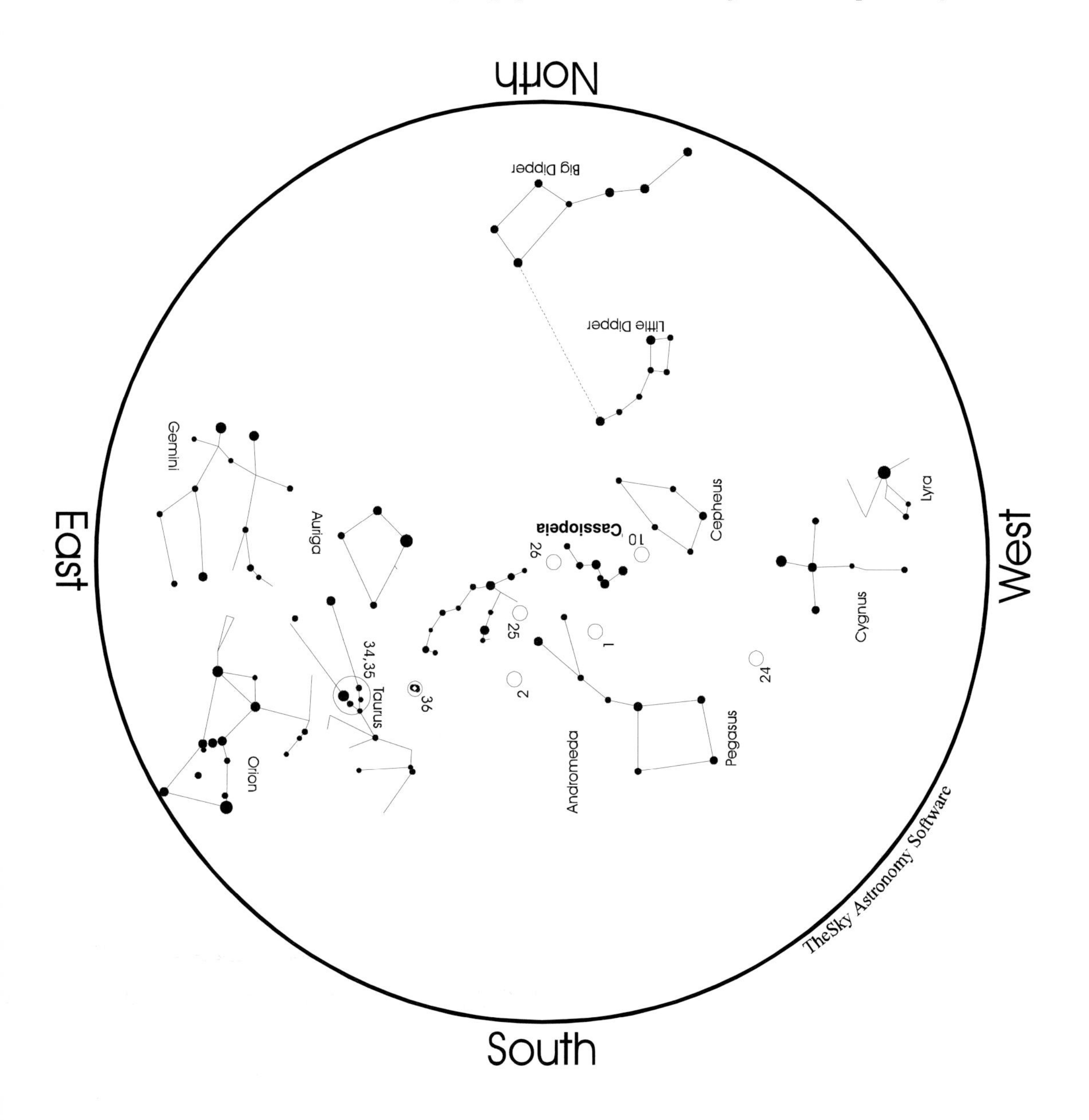

WHAT TO LOOK FOR

1. ANDROMEDA—M31, the Andromeda Galaxy—*spiral galaxy*

R.A.: 00h40.0m Dec.: +41°00'

Located just off Andromeda's belt, M31 will appear as a fuzzy, cigar-shaped object through your telescope. Because of its immense size you may not be able to see all of it in your field of view at one time. It is better to use a low-power, wide-angle eyepiece if you have one. If not, just move your telescope back and forth to enjoy the entire view. M31 is an excellent object for binoculars.

The Andromeda Galaxy is 2.2 million light-years away. It is 110,000 light-years in diameter and contains over 300 billion stars. For comparison, the Milky Way is thought to be only 100,000 light-years in diameter. Most other spiral galaxies are much smaller.

2. ANDROMEDA—NGC 752—*open cluster*

R.A.: 01h54.8m Dec.: +37°26'

Small telescopes have better luck with this cluster than large telescopes. This loose grouping of stars is scattered around an area too large for big telescopes to view at one time. Looking through a low-power, wide-angle lens, you will see a view scattered lightly with faint stars. NGC 752 contains just over 50 stars within a diameter of 17 light-years. The cluster is approximately 1,300 light-years away from us.

3. AURIGA—Capella—*yellow gold star*

R.A.: 05h13.0m Dec.: +45°57'

At first glance, you may only see a bright, white pinpoint of light through your telescope. But as your eyes adjust, you may see flashes of yellow or gold color. You are familiar with another yellow star, the Sun. Capella is the sixth brightest star in our sky. It lies 45 light-years away and is thought to be 160 times brighter than our Sun.

An interesting note: Although you can't see the individual stars through a telescope, astronomers have determined that Capella is a multiple star system. When you look at Capella, you aren't seeing the light from just one star. Instead, you are seeing the light from four stars orbiting each other. We can't resolve the four individual stars for two reasons: the great distance between us and Capella, and the relatively small distances between the four stars.

4. AURIGA—M36, M37 and M38—*open clusters*

M36	M37	M38
R.A.:	R.A.:	R.A.:
05h32.9m	05h49.0m	05h25.3m
Dec.: +34°07'	Dec.: +32°33'	Dec.: +35°48'

Lying to the south of Capella, these three individual clusters form a line through Auriga. Although they appear close together in the sky, it is thought that they formed separately. M36 lies between the two, slightly closer to M38. M36 and M38 can usually be seen in the same wide-angle field of view of a telescope. M36 is brighter and more condensed than M38. M37 is larger than the other two, and you will probably be able to make out more individual stars of this cluster.

M36: contains 60 stars; 4,100 light-years away; 14 light-years in diameter.

M37: contains 150 stars; 4,600 light-years away; 25 light-years in diameter.

M38: contains 100 stars; 4,200 light-years away; 25 light-years in diameter.

5. BOOTES—Arcturus—*yellow-gold star*
R.A.: 14h13.4m Dec.: +19°27'

Arcturus is the fourth brightest star in the night sky, so it is very easy to see. To make sure the star you are looking at is really Arcturus, follow the arc to Arcturus (see page 19). Arcturus is 37 light-years away and is 115 times brighter than our sun.

6. CANCER—M44—*open cluster*
R.A.: 8°37.5m Dec.: +19°52'

Also known as the Praesepe or the Beehive, this cluster can be seen with the naked eye as a small, faint cloud. However faint it may appear, M44 is a great cluster for binoculars and any small telescope with a wide-angle eyepiece. This cluster has several double stars within its 13 light-year diameter. M44 contains over 200 stars and is approximately 525 light-years away.

7. CANIS MAJOR—Sirius—*blue giant star*
R.A.: 06h43.0m Dec.: -16°39'

Sometimes called the "Dog Star," Sirius is the brightest star in the evening sky. You *can* see brighter objects at night, like the Moon, Jupiter, and Venus. But these objects are located in our solar system and are completely different from the stars. Sirius is 8.7 light-years from us and is over 23 times brighter than our Sun. Although Sirius is the brightest star in our night skies, it is not the closest. The Dog Star is actually the fifth closest star to us. The four stars that are closer to us are much fainter than Sirius.

8. CANIS MAJOR—M41—*open cluster*
R.A.: 06h44.9m Dec.: -20°42'

Just below (south of) Sirius, this open star cluster is one of the easier objects to find. On dark, clear nights M41 may be seen with the naked eye as a very faint, tiny cloud. Through a telescope, several bright stars can be seen. This cluster is loosely scattered over a 20-light-year diameter, is about 2,350 light-years distant, and may contain a total of 150 stars.

9. CANIS MAJOR—NGC 2362—*open cluster*
R.A.: 07h16.6m Dec.: -24°52'

NGC 2362 appears to be clustered around one of the stars of Canis Major. When you first look at this cluster, it may appear as if there is only a bright star with a hazy cloud surrounding it. Focus carefully and study the area around the star. You should be able to make out some of the 40 individual stars within the cluster. NGC 2362 is 4600 light-years away and about 8 light-years in diameter.

10. CASSIOPEIA—M52—*open cluster*
R.A.: 23h22.0m Dec.: +61°20'

Using the "W" shape of Cassiopeia can help you find this open cluster. Use the two stars that form the right side of the W (see the star chart). Of the two stars, start with the star at the bottom of the W and move to the top star. Then keep going until you have traveled about the same distance as between the two stars of the W. There you will find M52. This cluster contains about 200 stars within a diameter of 10 to 15 light-years. The cluster is 3,000 light-years distant.

11. COMA BERENICES—M53—*globular cluster*
R.A.: 13h10.5m Dec.: +18°26'

When looking for M53, don't look for a cluster of stars. Instead, look for a faint, round spot. This rather distinct round spot is actually a huge ball of over 50,000 stars. It appears as a faint, round spot because it is 65,000 light-years from us. M53 is located on the outer rim of our galaxy.

12. CYGNUS—Deneb—*supergiant star*
R.A.: 20h39.7m Dec.: +45°06'

Found in the tail of the Swan or the top of the Northern Cross, Deneb is the most distant of the brighter stars. Even though Deneb is 1,600 light-years away, it is still the 19th brightest star in the night sky. This is because Deneb is a supergiant star. Twenty-five times more massive than our Sun, Deneb is 60 times larger and 60,000 times brighter than our own star.

13. CYGNUS—Beta Cygni—*contrast double star*
R.A.: 19h28.7m Dec.: +27°51'

When looking at the star representing the head of the Swan or the base of the Northern Cross with your naked eye, you see only one star. A telescope will resolve that star into two separate stars of different colors. The brighter of the two has a yellow-gold color, while the second star has a blue tint. The two stars are 400 billion miles apart, or 100 times the distance between the Sun and Pluto. They are traveling around each other, but are moving too slowly for you to notice any movement. The star system is 410 light-years away.

14. CYGNUS AND BEYOND—*Milky Way*

Cygnus the Swan is flying along the hazy streak of light known as the Milky Way. Beginning at Deneb and scanning south towards the head of the Swan and on towards the constellations of Sagittarius and Scorpius, a small telescope can make out thousands of distant stars that form the spiral arms of our Galaxy. Dark streaks through the haze are huge clouds of dust that are blocking our view of the rest of our Galaxy.

15. GEMINI—M35—*open cluster*
R.A.: 06h05.7m Dec.: +24°20'

M35 is rather unusual for a star cluster because it lacks a central "glob" of stars. Rather, its 300 stars are evenly distributed throughout its 30 light-year diameter. When gazing at the cluster, you won't see all 300 stars. Instead you will see a hazy cloud without a brighter central region. You should be able to make out some brighter individual stars scattered throughout the cluster. M35 is 2,200 light-years distant.

16. HERCULES—M13—*globular cluster*
R.A.: 16h39.9m Dec.: +36°33'

Found in the keystone of Hercules, this bright, round, cloudy spot contains over 30,000 stars. Through a small telescope, though, you probably won't be able to make out any individual stars. The small spot you see through your telescope is over 160 light-years in diameter. It appears so small because M13 is over 24,000 light-years away from us.

17. HERCULES—M92—*globular cluster*
R.A.: 17h15.6m Dec.: +43°12'

If M92 were anyplace else in the sky, it would be considered a wonderful globular cluster. However, because it lies so close to the spectacular M13, it is often overlooked. M92 should be easily visible as a pale, fuzzy cloud, even though it is 35,000 light-years distant.

18. LEO—Regulus—*middle-aged star*
R.A.: 10h05.7m Dec.: +12°13'

Sometimes called "the heart of the Lion," Regulus can be found as the dot at the base of the backwards question mark that forms the head of Leo, the Lion. Regulus is 85 light-years away and about 5 times larger than our Sun. It may be 5 times bigger than our Sun, but it is an incredible 160 times brighter!

19. LEO—M65 and M66—*spiral galaxies*
R.A.: 11h17.0m Dec.: +13°20'

These two spiral galaxies can be found below the two stars that form the short end of the triangle of Leo and can usually be seen in the same wide-angle field of view. Don't expect to see beautiful spiral arms and the great detail you see in photographs. What you *are* going to see are two fuzzy blobs. After all, these galaxies are over 29 million light-years away from us.

Of the two, M66 is brighter. M65 may be easier to see, though, because it is larger. Keep in mind that these two faint, fuzzy blobs are completely separate from our own galaxy. Within each of those tiny blobs are billions of stars, planets, star clusters, nebulae, and much more. To contain all of those things, the galaxies have to be huge. M66 is 50,000 light-years in diameter and M65 is 60,000 light-years in size.

20. LYRA—Epsilon Lyrae—*double star*
R.A.: 18h42.7m Dec.: +39°37'

Epsilon Lyrae is a fancy name for the fifth brightest star in the constellation Lyra. It lies right next to the bright star, Vega. You can see Epsilon Lyrae without a telescope. But its unusual feature shows up only through a telescope.

Depending on your telescope, you will see anything ranging from something that looks like the number 8 to four individual stars. The 8 shape that you see in smaller telescopes results when the telescope is not quite able to resolve two separate stars. Larger telescopes are able to resolve Epsilon Lyrae into two individual stars. Even larger telescopes resolve these two stars into four stars. Epsilon Lyrae is actually a double-double star.

21. ORION—M42—the Orion Nebula—*diffuse nebula*
R.A.: 05h32.9m Dec.: -05°25'

M42 is the brightest diffuse nebula in our skies and can be seen with the naked eye as the fuzzy second star in Orion's sword. Even with a small telescope, you can make out details within the large, pale cloud of gas and dust. You can see these details because hot, young stars formed within this cloud cause the surrounding gases to heat up and glow. M42 is between 1,600 and 1,900 light-years away. The central portion of the cloud is approximately 30 light-years in diameter.

22. ORION—Rigel—*blue giant star*
R.A.: 05h12.1m Dec.: -08°15'

Rigel, a blazing blue-white giant star marking Orion's foot, is located 900 light-years away. It is almost 50 times larger than our Sun and produces energy at the incredible rate of 57,000 times the output of our Sun.

23. ORION—Betelgeuse—*red supergiant star*
R.A.: 05h52.5m Dec.: +07°24'

This red supergiant star marking Orion's shoulder is one of the largest stars known. It is also a variable star, shrinking and growing in size and brightness. Betelgeuse ranges in size from 550 to 920 times the size of our sun. To put things in perspective, imagine that Betelgeuse replaced our Sun. This red supergiant would completely swallow Mercury, Venus, Earth, and Mars, and its surface would almost reach Jupiter. Thankfully, we don't have to worry about that happening. Betelgeuse is 520 light-years away from us.

24. PEGASUS—M15—*globular cluster*
R.A.: 21h27.6m Dec.: +11°57'

Lying a little above and to the west of Pegasus' nose is a bright, compact globular cluster. Through a telescope, M15 appears as a small, round cloud. Because this cluster is so compact, there aren't many stars scattered around the edge of the cluster. M15 is over 39,000 light-years away from us and has a diameter of 130 light-years. Fifty to 100 thousand stars are crammed within this cluster.

25. PERSEUS—M34—*open cluster*
R.A.: 02h 38.8m Dec.: +42°34'

Located to the west of the second brightest star in Perseus, this open cluster will appear as a faint, irregular cloud. You should be able to make out a few individual stars within the cluster. M34 contains over 80 stars and is over 1,450 light-years away from us. As you look through your telescope at this small, irregular cloud, keep in mind that this "little" cluster is about 18 light-years in diameter (or one hundred six trillion, two hundred billion miles).

26. PERSEUS—h and χ (pronounced ky, as in sky) Persei—*double open cluster*
R.A.: 2h17.2m Dec.: +56°54'

NGC 869 and NGC 884 are other names for these two open clusters. Though they lie close enough together to be seen in the same field of view (binoculars or small telescopes) these clusters were not formed together. The closest cluster lies 7,400 light-years distant and contains almost 400 stars. The more distant cluster is 8,500 light-years away and contains about 300 stars.

27. SAGITTARIUS—M8—Lagoon Nebula—*diffuse nebula*
R.A.: 18h01.6m Dec.: -24°20'

Located to the west of the Teapot's spout, M8 is a beautiful diffuse nebula. Through a telescope, you should see an irregularly shaped cloud. The longer you observe, the more detail you can make out within the nebula's structure. You may be able to see a darker region within the nebula's center. This darker region is a band of dust that lies between us and the nebula. The nebula is approximately 5,150 light-years distant.

28. SAGITTARIUS—M21—*open cluster*
R.A.: 18h01.8m Dec.: -22°30'

Just north of M8 lies this open cluster of 50 stars. M21 has a very compact core of central stars with a few stars scattered around the edges. M21 is 2,200 light-years distant and 17 light-years in diameter.

29. SAGITTARIUS—*center of Milky Way Galaxy*

The center of our Galaxy can be found lying beyond the stars of the constellation Sagittarius. Though most of its intensity is blocked by huge dust clouds, the glow of distant stars is quite impressive. Scan the area to see what you discover.

30. SCORPIUS—Antares—*red supergiant star*
R.A.: 16h26.4m Dec.: -26°19'

Called "the heart of the Scorpion" for its reddish color in the night sky, Antares is one of the largest stars known. Its diameter is 700 times that of our sun, almost 600 million miles. Even though it is so large, it appears as only a pinpoint of light through a telescope. After all, it's over 520 light-years away. Antares appears red because of its relatively cool surface temperature, only 6,000°F (3,400°C) compared to our Sun's 10,000°F (5,500°C) temperature.

31. SCORPIUS—M4—*globular cluster*
R.A.: 16h20.6m Dec.: -26°24'

After you have found Antares, move your telescope to the west just a little and you'll find this beautiful globular cluster. Appearing as a large, circular, hazy patch, M4 is one of the closest and largest globular clusters. It contains well over 10,000 stars and is 5,700 light-years away.

32. SCORPIUS—M6 and M7—*open cluster*

These two open clusters lie between the tail of the Scorpion and the spout of the Teapot of Sagittarius.

M6 R.A.: 17h36.8m Dec.: -32°11'

You should be able to make out individual stars of this cluster forming patterns of lines within its irregular shape. M6 contains approximately 80 stars within a diameter of 20 light-years. Astronomers think this cluster is 1,500 light-years away.

M7 R.A.: 17h50.7m Dec.: -34°48'

Lying just below and to the east of M6, this loose grouping of stars contains at least 80 members. Although M7 appears to lie close to M6 in our skies, M6 is twice as far from us as M7. M7 is only 800 light-years away.

33. SCORPIUS—NGC 6231—*open cluster*
R.A.: 16h50.7s Dec.: -41°43'

Located within the curved tail of the Scorpion, this cluster consists of a central knot of 7 to 8 bright stars within a cluster of faint stars. The total group is within a diameter of 8 light-years and is 5,700 light-years distant.

34. TAURUS—Aldebaran—*red giant star*
R.A.: 04h33.0m Dec.: +16°25'

The brightest star in a V-shaped cluster of stars, Aldebaran is sometimes known as "the eye of the Bull." Aldebaran is a red giant star, which means you should see a reddish-orange tint to the starlight. This red giant star is 68 light-years away and is huge, almost 40 times larger than our Sun. Because it is bigger than our Sun, it is also brighter (almost 125 times brighter, to be exact).

35. TAURUS—the Hyades—*open cluster*

The V-shaped cluster of stars sometimes known as "the head of the Bull" is called the Hyades. The bright star Aldebaran highlights the top of one side of the V. Although very impressive in the night sky, the Hyades are really too large to see through a telescope. Try binoculars instead.

An interesting note: The star Aldebaran appears to lie within the Hyades cluster. In this case, appearances can be deceiving. Measurements show that the Hyades lie almost 130 light-years distant, over twice as far away from us as Aldebaran.

36. TAURUS—M45—the Pleiades—*open cluster*
R.A.: 03h43.9m Dec.: +23°58'

This small open cluster of stars often reminds people of a little dipper. On a clear night, without any visual aid, 6 to 7 stars can be seen within the group. Through a pair of binoculars almost 50 stars can be found. A small telescope will show several stars, but many people find the view through a telescope to be rather dull compared to what they can see with their own eyes. The Pleiades cover such a large area in the sky, a telescope can only see a small portion of the group at a time. M45 is 410 light-years away, and most of the stars in the cluster are within an area of 7 light-years in diameter.

37. URSA MAJOR (the Big Dipper)—Mizar/Alcor—*visual double star*

For these stars, don't use your telescope (at least at first). Mizar, the brighter of the two stars, is the second star in the handle of the Big Dipper. Alcor can be seen just above the brighter Mizar. Some stories claim that Alcor is riding piggyback on Mizar.

These two stars form what is called a visual double star system. That means they only *look* as if they are grouped together in the sky. Physically, they are not considered a group. If you try to observe them through a telescope, you will probably find that they are too far apart to be seen through the telescope at the same time.

37a. URSA MAJOR—Mizar—*binary star system*

R.A.: 13h21.9m Dec.: +55°11'

Now aim your telescope at Mizar, the second star in the handle of the Big Dipper. Mizar is an actual double star. Though a telescope, you will be able to see its companion. Keep in mind that this companion cannot be seen without a telescope. This star is not Alcor. Mizar is 88 light-years away. Its smaller companion lies the equivalent of five times the orbit of Pluto away from Mizar. The two stars travel very slowly around each other. You won't be able to see any movement through the telescope.

37b. URSA MAJOR—Alcor—*binary star system*

An interesting note: You won't be able to see this through your telescope, but Alcor is also a double star system. So, Mizar/Alcor appears to be a double star system without a telescope, but it really isn't. Yet Mizar and Alcor are each double stars. Confusing, isn't it?

38. URSA MAJOR—M81 and M82—*spiral galaxy and irregular galaxy*

M81	M82
R.A.: 09h51.5m	R.A.: 09h51.9m
Dec.: +69°18'	Dec.: +69°56'

You should be able to see these two galaxies in the same wide-angle field of view, but they are not going to be as easy to identify as a star or star cluster. True, each galaxy contains billions of stars. But these two groups of billions of stars are over 7 million light-years away. This huge distance tends to restrict the amount of detail you can see.

M81 is a spiral galaxy and is the larger of the two galaxies. Although it is larger than M82, it may be harder to see because it is rather spread out and there isn't a distinct shape to focus on. M82 is an irregular galaxy that looks like a hazy cigar.

Deep Sky Objects for Binoculars and Small Telescopes

January–February

Perseus—M34, h and χ Persei
Taurus—Aldebaran, Hyades, and Pleiades
Orion—M42, Rigel, Betelgeuse
Canis Major—Sirius, M41, NGC 2362
Gemini—M35
Auriga—Capella, M36, M37, M38

March–April

Leo—Regulus, M65, M66
Canis Major—Sirius, M41, NGC 2362
Cancer—M44
Orion—M42, Rigel, Betelgeuse
Auriga—Capella, M36, M37, M38

May–June

Leo—Regulus, M65, M66
Coma Berenices—M53
Bootes—Arcturus
Hercules—M13, M92
Ursa Major (Big Dipper)—Mizar/Alcor, M81, M82

July–August

Lyra—Epsilon Lyrae
Cygnus—Deneb, Beta Cygni, slowly scan across Milky Way
Hercules—M13, M92
Scorpius—Antares, M4, M6, M7, NGC 6231
Sagittarius—M8, M21, center of Milky Way

September–October

Pegasus—M15
Andromeda—M31, NGC 752
Cygnus—Deneb, Beta Cygni, slowly scan across Milky Way
Cassiopeia—M52
Sagittarius—M8, M21, center of Milky Way

November–December

Perseus—M34, h and χ Persei
Pegasus—M15
Andromeda—M31, NGC 752
Cassiopeia—M52
Taurus—Aldebaran, Hyades, and Pleiades